The Modern Therapeutic Radiographer

This study guidebook for 1st year undergraduate BSc therapeutic radiography/radiotherapy and oncology students introduces key academic and clinical subjects to help students successfully navigate the first year of their degree.

Breaking down key concepts in biology and physics, chapters cover essential areas of a radiotherapy course in a bitesize, easy-to-follow way, including key concepts such as oncological management, radiotherapy planning, dosimetry and radiobiology.

Information is presented in a clear and concise manner to help students understand exactly what they need to know, why they need to know it and how to apply it.

The book guides readers through essential knowledge demanded on placement and at university, written from an academic point of view, bringing together all relevant information in one handy guide. It enables readers to put academic knowledge into practice and provides real-world communication strategies, study tips and tricks, and topics such as academic support.

It is an ideal study guide for students and will be a useful reference for therapeutic radiography practice educators and academic therapeutic radiographers.

Key features:

- Written by academics with over 30 years of experience in academia and clinical practice.
- Tailored to 1st year level 4 undergraduate BSc therapeutic radiography students who have just started a degree program, looking for academic and clinical placement guidance and support.
- Written as a practically focused handbook, including a guide to pre-clinical placement, real-world communication strategies based on patient and staff interactions, study tips, and templates to complete.

Emma-Praer Breger is a lecturer at London South Bank University (LSBU) with over a decade's worth of experience in this field. She gained her Bachelor of Science (BSc) in Therapeutic Radiography from City University London. Emma is also undertaking a teaching apprenticeship at LSBU to further enhance her skills and knowledge as a lecturer. She has worked in a number of different hospitals, including Guys and St Thomas's and The Royal Sussex County Hospital.

Sophie Davis is a lecturer at London South Bank University (LSBU). She studied Biomedical Science at Oxford Brookes University, achieving an upper second-class honours degree in 2015. Sophie completed a master's with distinction in Therapeutic Radiography and Oncology at The University of the West of England in 2018. Sophie then went on to become a Band 5 Therapeutic Radiographer at The Royal Marsden Hospital in Sutton where she continued to work her way up to Senior Therapeutic Radiographer.

Rachael Williams is the course leader and senior lecturer for the Therapeutic Radiography programme at London South Bank University (LSBU). She qualified as a Therapeutic Radiographer in July 2010 from the University of the West of England, Bristol. She joined LSBU in 2016 having previously worked clinically at Guys and St Thomas' hospital and other London hospitals prior to that. She teaches across all the Therapeutic Radiography programmes and is passionate about radiotherapy planning & dosimetry and clinical practice.

Series in Medical Physics and Biomedical Engineering
Series Editors: Kwan-Hoong Ng, E. Russell Ritenour, and Slavik Tabakov

Biomedical Photonics for Diabetes Research
Edited by Andrey Dunaev and Valery Tuchin

Spectral Multi-detector Computed Tomography (sMDCT)
Data Acquisition, Image Formation, Quality Assessment and Contrast Enhancement
Xiangyang Tang

Monte Carlo in Heavy Charged Particle Therapy
New Challenges in Ion Therapy
Edited by Pablo Cirrone and Giada Petringa

Radiotheranostics – A Primer for Medical Physicists I
Physics, Chemistry, Biology and Clinical Applications
Edited by Cari Borrás and Michael Stabin

Image-guided Focused Ultrasound Therapy
Physics and Clinical Applications
Edited by Feng Wu, Gail ter Haar, and Ian Rivens

Clinical Insights for Image-Guidance Radiation Therapy
Prostate
Mike Kirby and Kerrie-Anne Calder

Essentials of Functional MRI
Patrick W. Stroman

Proton Therapy Physics, Third Edition
Edited by Harald Paganetti

Physiology, Biophysics, and Biomedical Engineering
Edited by Andrew Wood, Joseph Ciorciari and David Sly

An Introduction to Rehabilitation Engineering, Second Edition
Edited by Rory A. Cooper and Hisaichi Ohnabe

Radiotheranostics – A Primer for Medical Physicists II
Radiochemistry, Radiobiology, Dosimetry, Safety, Economics, and AI
Edited by Cari Borrás

The Modern Therapeutic Radiographer
An Introductory Guide
Emma-Praer Breger, Sophie Davis, and Rachael Williams

For more information about this series, please visit: www.routledge.com/Series-in-Medical-Physics-and-Biomedical-Engineering/book-series/CHMEPHBIOENG

The Modern Therapeutic Radiographer
An Introductory Guide

Emma-Praer Breger, Sophie Davis
and Rachael Williams

CRC Press
Taylor & Francis Group
Boca Raton London New York

CRC Press is an imprint of the
Taylor & Francis Group, an **informa** business

Designed cover image: Belinda Breger

First edition published 2027
by CRC Press
2385 NW Executive Center Drive, Suite 320, Boca Raton FL 33431

and by CRC Press
4 Park Square, Milton Park, Abingdon, Oxon, OX14 4RN

CRC Press is an imprint of Taylor & Francis Group, LLC

ISBN: 978-1-041-07173-0 (hbk)
ISBN: 978-1-041-06244-8 (pbk)
ISBN: 978-1-003-63918-3 (ebk)

DOI: 10.1201/9781003639183

Typeset in Minion
by KnowledgeWorks Global Ltd.

This book is dedicated to all our family and friends who have supported us throughout our studies and careers.

Contents

About the Series

The *Series in Medical Physics and Biomedical Engineering* describes the applications of physical sciences, engineering, and mathematics in medicine and clinical research.

The series seeks (but is not restricted to) publications in the following topics:

- Artificial organs
- Assistive technology
- Bioinformatics
- Bioinstrumentation
- Biomaterials
- Biomechanics
- Biomedical engineering
- Clinical engineering
- Imaging
- Implants
- Medical computing and mathematics
- Medical/surgical devices
- Patient monitoring
- Physiological measurement
- Prosthetics

- Radiation protection, health physics, and dosimetry
- Regulatory issues
- Rehabilitation engineering
- Sports medicine
- Systems physiology
- Telemedicine
- Tissue engineering
- Treatment

The *Series in Medical Physics and Biomedical Engineering* is an international series that meets the need for up-to-date texts in this rapidly developing field. Books in the series range in level from introductory graduate textbooks and practical handbooks to more advanced expositions of current research.

The *Series in Medical Physics and Biomedical Engineering* is the official book series of the International Organization for Medical Physics.

The International Organization for Medical Physics

The International Organization for Medical Physics (IOMP) represents over 18,000 medical physicists worldwide and has a membership of 80 national and 6 regional organisations, together with a number of corporate members. Individual medical physicists of all national member organisations are also automatically members. The mission of IOMP is to advance medical physics practice worldwide by disseminating scientific and technical information, fostering the educational and professional development of medical physics and promoting the highest quality medical physics services for patients.

A World Congress on Medical Physics and Biomedical Engineering is held every three years in cooperation with International Federation for Medical and Biological Engineering (IFMBE) and International Union for Physics and Engineering Sciences in Medicine (IUPESM). A regionally based international conference, the International Congress of Medical Physics (ICMP) is held between world congresses. IOMP also sponsors international conferences, workshops and courses.

The IOMP has several programmes to assist medical physicists in developing countries. The joint IOMP Library Programme supports 75 active libraries in 43 developing countries, and the Used Equipment Programme coordinates equipment donations. The Travel Assistance Programme provides a limited number of grants to enable physicists to attend the world congresses.

IOMP co-sponsors the *Journal of Applied Clinical Medical Physics*. The IOMP publishes, twice a year, an electronic bulletin, *Medical Physics*

World. IOMP also publishes e-Zine, an electronic newsletter about six times a year. IOMP has an agreement with Taylor & Francis for the publication of the *Medical Physics and Biomedical Engineering* series of textbooks. IOMP members receive a discount.

IOMP collaborates with international organisations, such as the World Health Organisations (WHO), the International Atomic Energy Agency (IAEA) and other international professional bodies such as the International Radiation Protection Association (IRPA) and the International Commission on Radiological Protection (ICRP), to promote the development of medical physics and the safe use of radiation and medical devices.

Guidance on education, training and professional development of medical physicists is issued by IOMP, which is collaborating with other professional organisations in development of a professional certification system for medical physicists that can be implemented on a global basis.

The IOMP website (www.iomp.org) contains information on all the activities of the IOMP, policy statements 1 and 2 and the 'IOMP: Review and Way Forward' which outlines all the activities of IOMP and plans for the future.

Preface

During an induction day, we spoke of the challenges that first-year students can face. As the discussion progressed, we observed that quite often students become overwhelmed by the barrage of information presented to them before they begin their placements. In addition, the stark contrast between university and clinical placement can be jarring. The change of pace, the conversations that occur and applying theory to practice quickly becomes bewildering, and it can be difficult to know where and who to turn to for guidance. Similarly, looking back to when we were first-year students, we would have greatly appreciated an easy-to-read guide regarding the basics of the academic aspects of radiotherapy and unravel the complexities of a radiotherapy department, including what was to be expected of us as students. It occurred to us that a potential solution would be a handbook that bridged the gap between university and clinical placement. Thus, the concept of *The Modern Therapeutic Radiographer: An Introductory Guide* was born over coffee on a chilly induction day, full of anticipation for the year ahead.

CHAPTER 1

Introduction

WHO ARE THERAPEUTIC RADIOGRAPHERS

Therapeutic radiographers are allied health professionals (AHP) and core members of a team dedicated to helping people with cancer. To practice as a therapeutic radiographer in the United Kingdom (UK), you are required to be registered with the Health and Care Professions Council (HCPC). Therapeutic radiographers are responsible for delivering safe and effective radiotherapy treatment, whilst ensuring the highest standard of patient-centred care.

WHAT IS RADIOTHERAPY?

Radiotherapy treatment is the use of ionising radiation to kill or slow down the production of tumour cells, whilst sparing healthy tissues that surround the tumour or tumour bed, in both malignant (cancerous) and benign (non-cancerous) conditions.

INTERNAL AND EXTERNAL RADIOTHERAPY

Both benign and malignant diseases can be treated with either internal or external radiotherapy.

External beam radiotherapy (EBRT) is delivered from outside of the body. EBRT can be used to treat any part of the body, including the skin. It is the most common form of radiotherapy and can be further categorised into different treatment techniques such as intensity modulated radiotherapy (IMRT) or volumetric modulated arc therapy (VMAT).

DOI: 10.1201/9781003639183-1

Internal radiotherapy is delivered from inside the body and is known as brachytherapy. Unlike EBRT, brachytherapy can only treat certain parts of the body such as the cervix or prostate.

TREATMENT MODALITIES

Radiotherapy is one of the fastest-growing sectors in cancer treatment. Through the implementation of technological advancements, the production of treatment modalities has moved on significantly from the 1950s. Modern machines no longer rely on a live radioactive source; instead, they employ particle acceleration technology to generate high-energy radiation in the form of photons, protons or electrons.

Linear accelerators, commonly abbreviated to linacs are the name of the most common treatment modality. It is important to note that most of the machines in the UK are made by either Elekta or Varian. These are the names of the medical companies that produce linear accelerators. Each company produces linear accelerators that work a similar same way; however, they may look different or have specific features only that company produces.

THE ROLE OF A STUDENT THERAPEUTIC RADIOGRAPHER IN THE RADIOTHERAPY DEPARTMENT

Therapeutic radiographers play a crucial role in the treatment of cancer patients and supporting patients through their radiotherapy treatment. As a student therapeutic radiographer, you are joining the frontline battle against cancer and will quickly become a vital part of any therapeutic radiography team.

As a student therapeutic radiographer, you will be based in a clinical radiotherapy department, which is usually located in the basement of the hospital, due to radiation safety laws. You will be placed in a team in a specific area of radiotherapy practice, whilst your main role on placement is learning and progressing through your studies, you have many other important roles. To understand these roles, it is important to break down the following:

1. The day-to-day role of a therapeutic radiographer
   A therapeutic radiographer uses ionising radiation to treat cancers and tumours. This is achieved by using technical equipment such as a linear accelerator to precisely target the cancer within the patient. A therapeutic radiographer is responsible for ensuring the machine

is ready for use every morning, organising and planning the patients for the day, ensuring patients are ready for their treatment, supporting patients' emotions and physical needs and safely delivering radiotherapy treatment.

2. The types of patients that are treated in a radiotherapy department
   The majority of patients who enter a radiotherapy department are having treatment for cancer. Some patients will be very unwell and/or at the end of their life, whilst others can be well. This depends on a multitude of things such as where the cancer is, where the cancer has spread to and any other treatments the patient may be undergoing. All patients will be at a different part of their cancer journey and will need a varying amount of support. It is important to note that patients can be scared, upset and in pain. Patient care is a large part of a therapeutic radiographer's role.

   Student therapeutic radiographers play a pivotal role in the smooth operation of the day-to-day running of the machine they are allocated to. Outlined below are the roles that student radiographers can take on.

(Note these roles are suggestions and should only be carried out when the student feels comfortable and competent, under sufficient supervision and in line with the students' learning objectives.)

Patient preparation and queue management:

1. Calling patients from the waiting room and ensuring they are changed into a gown in a timely fashion.
2. Ensuring patients have done their preparation for treatment such as drinking water to fill their bladder or fasting if required for their treatment.

Machine preparation:

1. Setting the room up for the next patient, such as selecting the correct immobilisation and cleaning the room from the previous patient.
2. Ensuring the correct equipment is available for the patient.
3. Resetting the machine, i.e. placing the gantry and collimators back to the centre.

Emotional support and patient care:

1. Students do not have the same time constraints as radiographers and are often best placed to be able to speak at length with patients. This can mean that students can form rapport with patients and support them through their treatment, referring them to a supervising radiographer when needed.
2. Some patients will need more support than others, both emotionally and physically. Students can assist these patients in several ways such as ensuring they know when their next appointment is and escorting them back to the hospital transport service.
3. General patient care will be one of your main roles to help ensure outstanding and quality patient-centred care is always delivered to each individual patient.

General support:

1. Often, there are many little jobs to do to ensure the general smooth running of the day. These include communicating with other healthcare professionals, returning equipment, ensuring patients medications are collected and ensuring patients have the correct appointment time. Students can take on some of these jobs to support the team and their interprofessional learning objectives.

Whilst the role of a student in the radiotherapy department is wide-ranging, you should always keep focus on your learning as that is what you are there to do. Use your duties to develop your skills and knowledge. When undertaking any of the roles discussed, try to link them back to your objectives and think about how you can enhance your learning. For example, if you are emotionally supporting a patient, this could be linked to a communication objective or supporting your ability to be compassionate and empathetic.

As learning is your main goal whilst on placement, developing your technical and non-technical skills is crucial to successful learning. Below is a list of non-technical skills that are essential to develop as a future therapeutic radiographer. Think about how you can achieve these in your role as a student therapeutic radiographer.

- Compassion (defined as a strong feeling of sympathy for others who are suffering, coupled with a desire to help alleviate that suffering)

- Empathy (defined as the ability to emotionally understand what others feel and be able to put yourself in their shoes)
- Giving information clearly and concisely
- Identifying and tackling barriers to communication. For example, if a patient requires an interpreter
- Receiving information from health care professionals, patients, family members and carers
- Maintaining standards such as the HCPC code of conduct, performance and ethics
- Managing workload
- Planning and prioritising
- Supporting others
- Recognising and understanding your strengths and weaknesses

A student therapeutic radiographer must adhere to the HCPC Standards of Conduct, Performance and Ethics for Students and base their roles around this (HCPC 2024).

Reminder: Always seek support from your supervising radiographer or practice educator and only work within the limits of your knowledge and understanding (HCPC 2024). Your main role is to learn, and all other roles are to support this learning. Remember you are a valuable member of any team.

REFERENCE

Health and Care Professions Council (HCPC) 2024. Guidance on conduct and ethics for students. Guidance on conduct and ethics for students | The HCPC (accessed 18 December 2025).

CHAPTER 2

Anatomy and Biological Sciences

INTRODUCTION TO BIOLOGICAL SCIENCES

This chapter will cover simple biological science and radiobiological principles. To understand how radiotherapy works at a cellular level, it is important to first understand the structure and function of cells and how they divide to form new cells. This chapter will also look at how neoplasms develop at a cellular level and how radiobiology is applied in radiation safety. Finally, you will learn about anatomical planes, directions and relevant anatomy.

THE CELL: STRUCTURE AND FUNCTION

Cells are the body's basic building blocks. They are the smallest functional units of the body. When cells are grouped together, they form tissues, each of which has a specialised function. An example of this would be muscle tissue. Different tissues are then grouped together to form organs such as the lungs, heart and liver. Organs form systems which have functions within the body such as the digestive system or reproductive system.

Eukaryotic cells are found in multicellular organisms such as humans and contain a nucleus.

The human body develops from a single cell called the zygote, which results from the fusion of the ovum (female egg cell) and the spermatozoon (male sex cell). These reproductive cells are known as germ cells. All cells that are not reproductive cells are called somatic cells. As the foetus

DOI: 10.1201/9781003639183-2

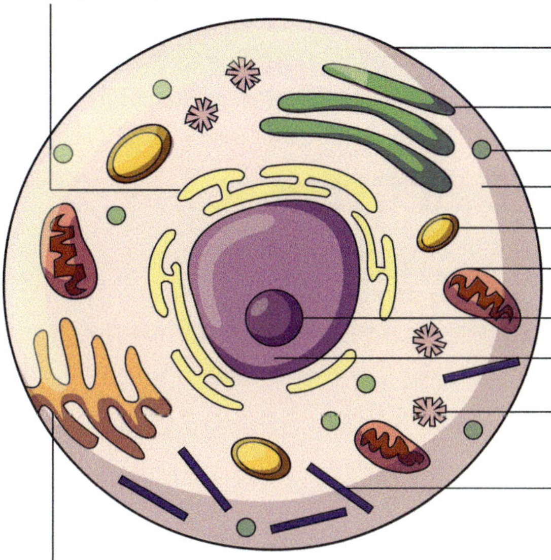

FIGURE 2.1 An animal eukaryotic cell. (Structure Animal Cell Labeled Parts Biology Stock Vector [Royalty Free] 769998433 | Shutterstock.)

develops, cells begin to differentiate into various types with distinct structures and specialised functions, such as blood cells, immune cells and nerve cells.

A cell consists of a plasma membrane that contains organelles, as seen in Figure 2.1. Organelles are specialised structures within the cell that carry out important functions and are suspended in a watery fluid within the cell known as the cytoplasm. Examples of organelles include ribosomes and endoplasmic reticulum.

THE CELL CYCLE

Cells undergo cell division to replace dead, worn out or damaged cells. The regularity at which cells divide is normally determined by the type of cell or tissue. The cell cycle is carefully maintained by checkpoints to ensure that only healthy cells are dividing. Somatic (diploid) cells divide by mitosis, while germ cells undergo meiosis to produce haploid gametes. Diploid cells contain 46 chromosomes (in humans) and undergo mitosis, in which the cell divides once to produce two genetically identical diploid cells. In contrast, diploid germ cells undergo meiosis, which involves two cell divisions and results in four genetically non-identical haploid sex cells, each containing 23 chromosomes. The next section will focus on mitosis. There are two main parts of the cell cycle: interphase, which is cell growth and preparation for mitosis and mitosis itself where the cell divides.

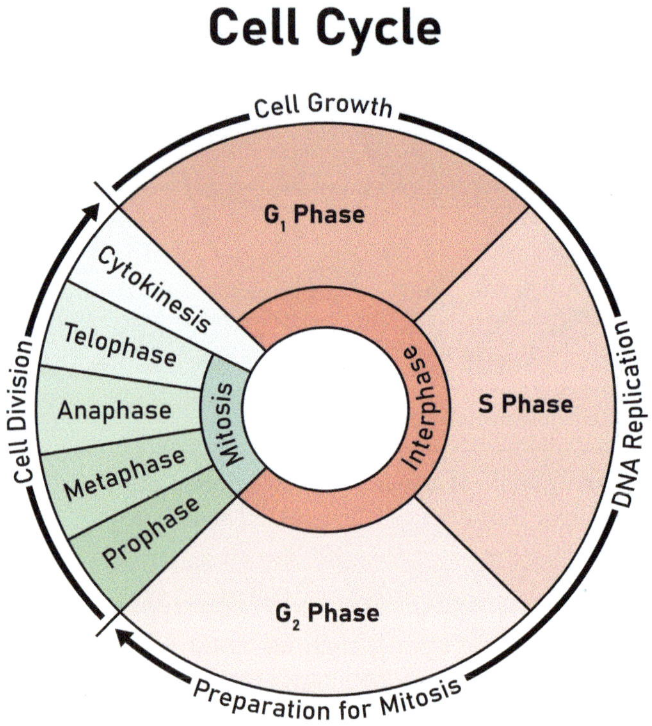

FIGURE 2.2 The cell cycle of a diploid cell. (https://www.shutterstock.com/image-vector/cell-cycle-division-scientific-design-vector-2389038979.)

Interphase

Interphase makes up 90% of the cell cycle and is the longest phase of the cell cycle. Interphase is a phase of cell growth and copying of the chromosomes. Interphase is split into different stages.

- Gap 1 (G1)
  This is the longest stage, but it does vary in duration. During the first gap phase, cells grow in size and volume. Organelles such as ribosomes and mitochondria are produced within the cell in this stage. Protein and mRNA (messenger ribonucleic acid) are synthesised ready

for S phase. If the conditions of the cell are not suitable for replication the cell will enter G0 (resting phase).

- Synthesis (S)
  DNA (deoxyribonucleic acid) replication occurs during this phase ready for mitosis. Each chromosome is duplicated, producing two identical sister chromatids. Histone proteins are synthesised, these proteins help to pack the new chromosomes.

- Gap 2 (G2)
  This is the final stage of interphase where the cell prepares for division by ensuring DNA integrity and producing necessary components for mitosis.

  This includes the cell continuing to grow in size, additional organelles are synthesised, tubulin is formed and the cell checks for any damage in the newly synthesised DNA, if any of the DNA is damaged and cannot be repaired, the cell may undergo apoptosis (cell death), otherwise the cell is ready to enter mitosis.

Mitosis (M Phase)

Cell division occurs in the M phase; there are four distinct stages which are visible by light microscopy.

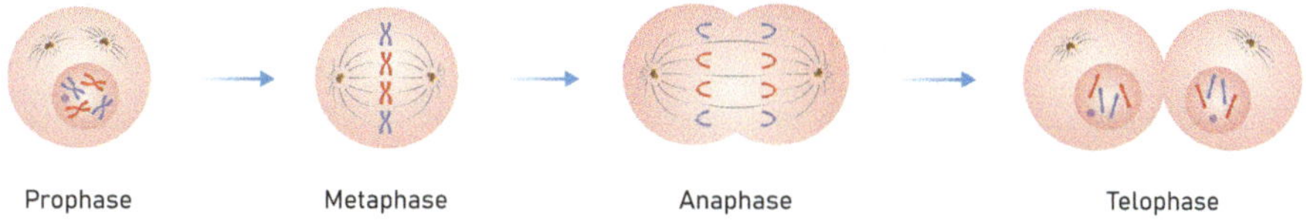

FIGURE 2.3 Mitosis phase of cell division. (https://www.shutterstock.com/image-vector/mitosis-phases-scientific-design-vector-illustration-2388171241.)

- Prophase
  During prophase, the loosely coiled, replicated chromatin becomes tightly coiled and is paired with its copy, condensing into chromosomes. Each chromosome consists of two sister chromatids joined at the centromere. The mitotic apparatus appears; this consists of two centrioles separated by the mitotic spindle, which is formed from microtubules. The centrioles then migrate, one to each end of the cell. The nuclear envelope breaks down, which allows the chromosomes to move freely ready for the next stage.

- Metaphase
  Chromosomes, each made of two chromatids line up in the middle of the cell attached by their centromeres.

- Anaphase
  The centromeres separate, the microtubules contract, causing the spindle fibres to shorten, and the replicated chromosomes are moved to opposite poles of the cell.

- Telophase
  The mitotic spindle breaks down as it is no longer needed. The chromosomes uncoil and the nuclear membrane reforms around each set of chromosomes creating two nuclei.

  After telophase, cytokinesis occurs, where the two new identical daughter cells are formed, by the splitting of the cytosol, intracellular organelles and plasma membrane.

DEVELOPMENT OF NEOPLASMS

Carcinogenesis is the development of cancer. There are three main stages:

1. Initiation
   The first stage of carcinogenesis involves DNA being damaged, which can be a result of exposure to carcinogens such as chemicals, viruses or radiation. DNA can also be damaged as a result of gene mutations in critical genes such as tumour suppressor genes (e.g. TP53). The damage disrupts the normal functioning of the gene leading to the next stages of carcinogenesis.

2. Promotion
   During this stage, cellular changes occur as 'promoters' stimulate mutated cells to proliferate. 'Promoters' such as hormones, inflammation or growth factors play a crucial role in facilitating this proliferation. These changes pave the way for progression into tumour cells.

3. Progression
   Mutated cells or tumour cells now undergo genetic changes to make them more aggressive. These changes can include angiogenesis (formation of new blood vessels) and evading growth suppressors, leading to uncontrolled cell growth. Tumour cells can gain the ability to invade nearby cells or travel around the body via the blood or lymph, known as metastasis.

RADIOBIOLOGY

Radiobiology is the study of the effects of ionising radiation on living organisms, particularly at a cellular and molecular level. This area of science has been used to help understand the effects on populations in atomic bomb zones and in the Chernobyl disaster as well as helping to shape radiation safety laws and protocols.

Ionising radiation causes damage to cells leading to cellular injury. The damage to cells depends on the type of radiation, radiation dose and the length of exposure. Once a person has been exposed to ionising radiation via radiotherapy the following occurs:

Ionising radiation interacts with organic tissue (cells in the body)

⇩

Ionisation occurs

⇩

Electrons are liberated

⇩

Further ionisation and excitation occur along the track

⇩

Damage is caused to organic molecules

Direct and Indirect Action

Radiation damage occurs either via direct action or indirect action. Direct action assumes the ionisation occurs directly with the cell's essential molecules (DNA, proteins, biomolecules). Indirect action occurs when radiation interacts with a water molecule in the body. This action produces free radicals which then go on to interact with the DNA causing cellular injury. Both actions can result in a single-strand break (SSB) or double-strand break (DSB) in the DNA helix. DSBs are harder to repair and more likely to cause cell death. However, SSB are more likely to occur when using MV (megavoltage) photons; this is why a fractionated dose of radiotherapy is utilised to continue to cause cellular injury and eventually causing cell death of tumour cells.

Linear Energy Transfer

MV photons used in radiotherapy are classed as low linear energy transfer (low-LET) radiation. LET is the average energy deposited per unit path length along the track of an ionising particle. It determines the biological consequence of a radiation exposure. MV photons are sparsely ionising and cause cellular injury by indirect action via free radical formation. An example of a high LET radiation is alpha radiation which are densely ionising particles. Each type of radiation is given a radiation weighting factor (WR) that accounts for the relative biological effectiveness (RBE); this is used in radiation protection (see below).

The 5R's of Radiobiology

Radiotherapy exploits key radiobiological principles in order to effectively treat cancer. These are described as the 5R's of radiobiology and are the principles of fractionated radiotherapy.

1. Repair
   Cancer cells and healthy cells can repair damage to their DNA following radiation exposure. Repair of cells is dependent on dose. Therefore, a higher dose is favourable; however, healthy tissues are a limiting factor, as tumour kill needs to be balanced with normal tissue complications. Some normal tissues repair more effectively than tumour cells allowing fractionated radiotherapy to be effective. There should be at least six hours between fractions to allow healthy cells to repair.
2. Reassortment
   Cells at different points in the cell cycle have different sensitivities to radiation. For example, cells in the G1 or S phase are more radioresistant and cells in the M phase are more radiosensitive. Fractionating radiotherapy allows cells to 'reassort' through the cell cycle, meaning cells that were originally resistant would move to a radiosensitive phase therefore improving tumour kill. At least a six-hour time interval between fractions is often utilised for reassortment to occur.
3. Repopulation (redistribution)
   Repopulation refers to the growing back of cells after irradiation. In some cases, normal tissues repopulate faster than tumour cells, protecting them from excessive radiation damage. However, in tumour populations, accelerated repopulation can occur due to the injured cell population being stimulated to cycle and divide more

rapidly. Therefore, longer and/or prolonged treatment times can be detrimental as tumour control cannot be achieved. Fractionation schedules can be manipulated to overcome this. Hyperfractionation delivers smaller doses per fraction (typically <1.8 Gy) more than once daily, with at least six hours between treatments to allow normal tissue repair. However, the overall treatment times stay the same. Hypofractionation is another option; this schedule delivers a larger dose per fraction; overall treatment time is reduced; fractions can be delivered on alternate days.

Repopulation is an important consideration if a patient's treatment is interrupted, which can be caused by machine breakdowns, patient illness or bank holidays.

The Royal College of Radiologists has defined three categories of patients based on tumour cell repopulation. These categories outline how a patient should be prioritised if there is an unplanned treatment interruption.

TABLE 2.1 RCR Categories of Patients Based on Tumour Cell Repopulation

Category	Definition	Tumour Type Examples
1	Patients with the tumour types for which there is evidence that prolongation of treatment affects outcome, and who are being treated radically with curative intent.	Non-small cell lung carcinoma Squamous cell carcinomas
2	Patients with slower-growing tumour types, who are being treated radically, where interruptions in radiotherapy leading to an extension of overall treatment time of more than five days are detrimental to both local control and survival.	Transitional cell carcinoma of the bladder Carcinoma of the prostate
3	Patients being treated palliatively	Metastatic bony tumour

Source: RCR (2019).

4. Reoxygenation

   The presence of oxygen enhances radiation induced DNA damage. Indirect action (as detailed above) is enhanced by oxygen as the oxygen reacts with the DNA free radicals, preventing DNA repair. However, many tumours are fast-growing and outgrow their blood supply making hypoxic (low levels of oxygen) cells within the tumour radioresistant (meaning they are resistant to the effects of radiation). Fractionated radiotherapy allows time for hypoxic cells to become reoxygenated.

5. Radiosensitivity
   The term radiosensitivity describes how likely the cell is to be damaged by radiation. Undifferentiated cells and highly proliferative cells are more radiosensitive. Resistance to radiation increases with cell maturity. High sensitivity tissues include gastric mucosa, oesophageal epithelium and primitive blood cells. Low-sensitivity tissues include nerve and muscle tissues. Dose and overall treatment time should be adjusted to account for the tissues within the treatment field.

Stochastic and Deterministic Effects

The effects of radiation on the body manifest in side effects. However, radiobiological effects can be categorised into two categories stochastic effects or deterministic effects.

Stochastic effects are random in nature, and the likeliness of a stochastic effect occurring increases with increasing dose. An example of a stochastic effect is the development of a secondary cancer; there is no known dose at which a secondary cancer will develop, but the chance increases with increasing dose. Each tissue has a weighting factor which is a relative measure of risk of stochastic effects that might result from irradiation of that tissue.

A determinist effect is an effect that occurs at a known threshold dose and the severity of the effect increases with increasing dose. An example of this is erythema of the skin; we know this will occur after a certain dose is absorbed, and the severity of the skin reaction increases with the increasing dose over this threshold.

Radiobiology in Radiation Safety

As mentioned above, radiobiology is used to determine safe practices for radiotherapy and help shape radiation protection legislation. Examples of this are in the Ionising Radiation Regulations 2017 (IRR17) (Legislation.gov.uk 2017), a safe annual dose limit has been determined for ionising radiation workers aged over 18, aged 16–18 and the public. The concept of as low as reasonably practicable (ALARP) was also developed to help reduce unnecessary radiation exposure.

Another example of how radiobiological research has helped to shape radiation protection is in the building of the radiotherapy bunker and the shielding that is used to line the bunker (thick concrete), along with the distance between the linac and the radiographers to ensure minimal radiation exposure to radiographers and members of the public.

There are three definitions of dose in radiological protection (radiation protection)

TABLE 2.2 ICRP 2019 Definitions of Dose in Radiobiological Protection

Dose	Explanation	Factors Needed	Units Expressed In
Absorbed dose	A measurable physical quantity, it is a measurement of how much energy is deposited within a mass such as tissue.	–	mGy (Milligrays)
Equivalent dose	Used for radiological protection purposes, calculated for an individual organ, using absorbed dose and adjusted to account for the effectiveness of the type of radiation.	Radiation weighting factor (WR)	millisieverts (mSv)
Effective dose	Used for radiological protection purposes, effective dose is calculated for the whole body. It is the addition of equivalent doses to all organs, each adjusted to account for the sensitivity of the organ to radiation.	Tissue weighting factor (WT)	millisieverts (mSv)

Source: Yonekura (2019).

Radiographers are often monitored by personal dosimeters to ensure a safe level of radiation is not exceeded. The average annual exposure from background levels of radiation in the UK (all sources) is around 3 mSv a year. On average, an employee working with ionising radiation receives an occupational exposure dose of less than 1 mSv a year, in contrast, on average, the annual occupational exposures for long-haul cabin crew and airline pilots are about 3mSv/year. Radiation dose is a measure of the amount of exposure to radiation.

BODY PLANES

Anatomical terms are used to describe body regions, movement and relationships between structures. Using correct anatomical terms in radiotherapy is important as terms are precise and standardised, meaning communication between radiographers can be clear and concise.

The anatomical position describes the starting point for any description of the human body. The body is in the upright position with the head facing forward, the arms at the sides with the palms of the hands facing forward and the feet together. In radiotherapy, the patients are often supine or prone.

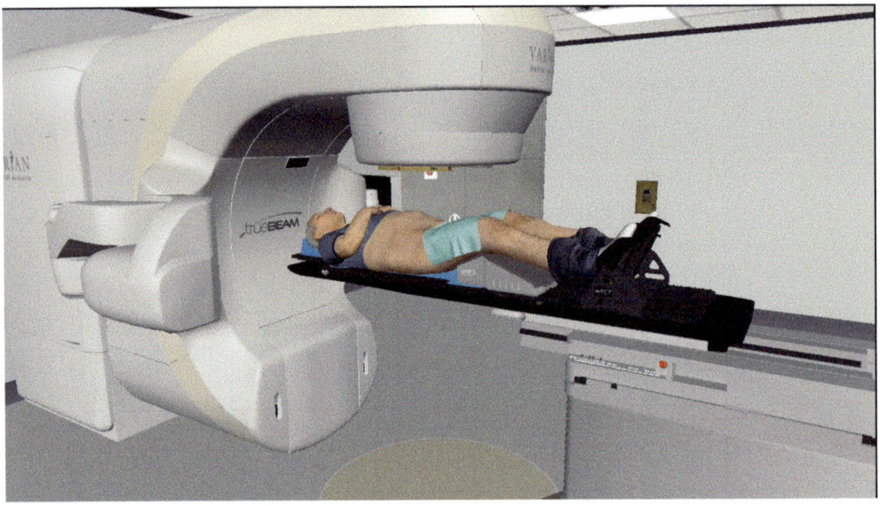

FIGURE 2.4 Patient supine on a linear accelerator. (With Permission from Vertual.)

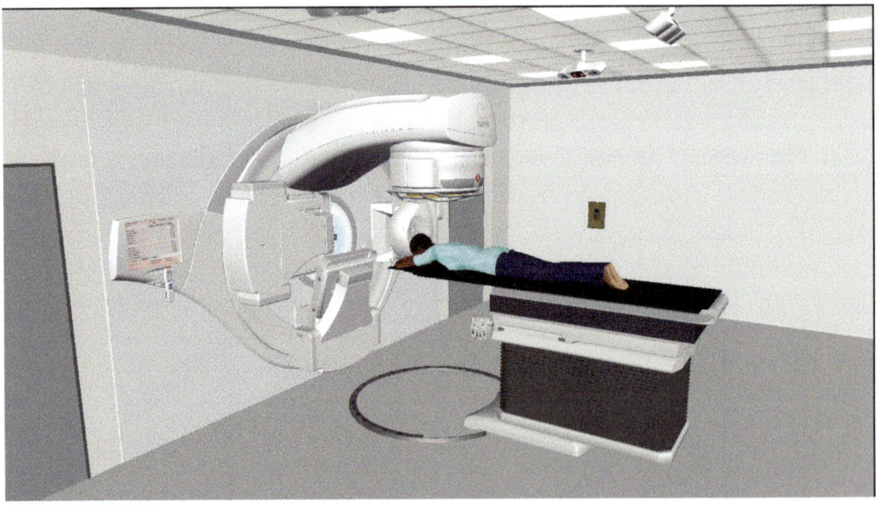

FIGURE 2.5 Patient prone on a linear accelerator. (With Permission from Vertual.)

Directional terms describe positions of structures relative to each other. An example for how these terms are utilised in radiotherapy is using them during patient set up to describe the position of the lasers in relation to the patient's permanent mark. See Table 2.3 for a description of the terms used.

TABLE 2.3 Directional Terms

Term	Meaning
Superior	Nearer to the head. *The mouth is superior to the shoulders.*
Inferior	Nearer to the feet. *The shin is inferior to the knee.*
Anterior	Nearer to the front of the body. *The sternum is anterior to the spine.*
Posterior	Nearer to the back of the body. *The spine is posterior to the sternum.*
Distal	Further from a point of attachment of a limb, or origin of a body part. *The fibula is distal to the femur.*
Proximal	Nearer to a point of attachment of a limb, or origin of a body part. *The femur is proximal to the fibula.*
Medial	Structure nearer to midline. *Patient's medial tattoo is placed on xiphisternum.*
Lateral	Structure further from midline or at the side of the body. *Patient's lateral tattoos are placed on the side of their body.*
Superficial	Superficial structures are closer to the skin. *The skin is superficial to the muscle.*
Deep	Far from the body surface. *The muscle is deep to the skin.*

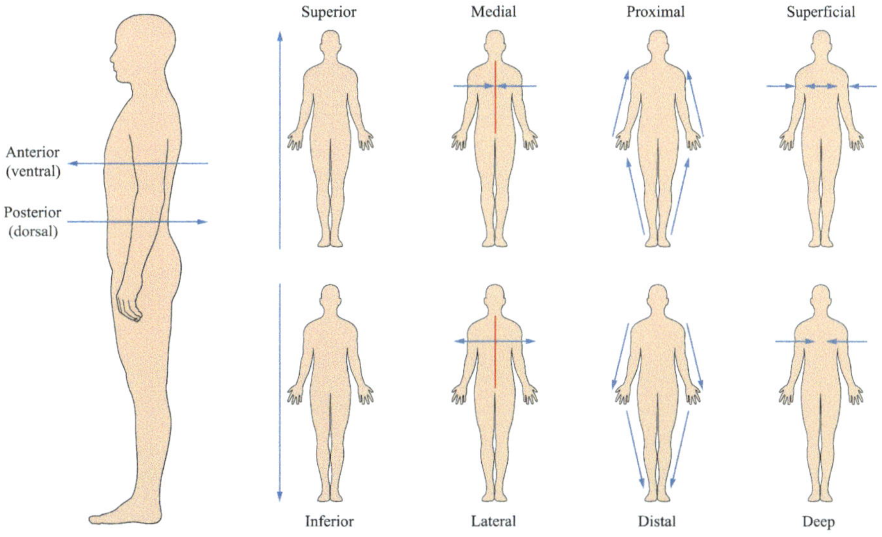

FIGURE 2.6 Directional terms. (https://www.shutterstock.com/image-vector/anatomical-directional-terms-illustration-man-body-1645350637.)

There are three body planes which lie at right angles to each other. These divide the body into sections and in radiotherapy are most commonly used when visualising 2D or 3D X-ray images taken of the patient.

Coronal (frontal) divides the body longitudinally into front (anterior) and back (posterior) planes.

Sagittal (medial) divides the body longitudinally through the midline into left and right halves.

Transverse (axial) divides the body horizontally into upper (superior) and lower (inferior) parts.

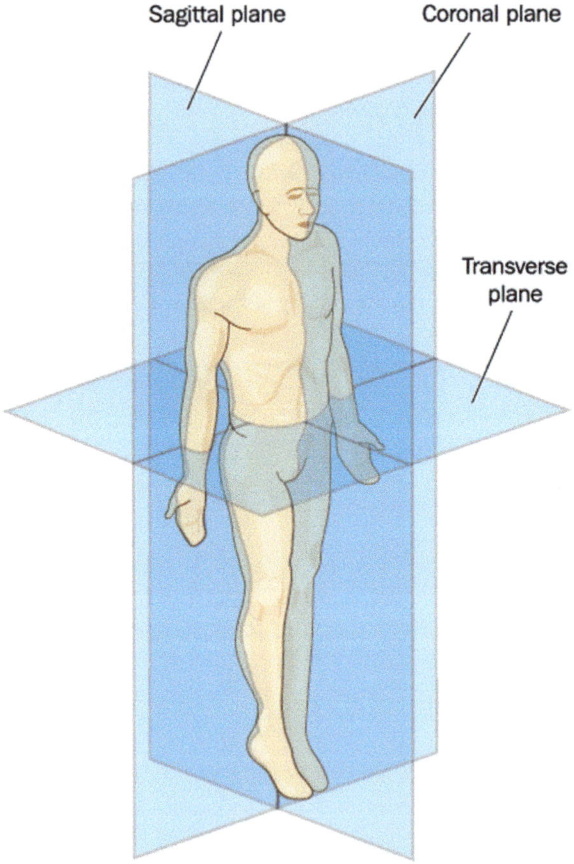

FIGURE 2.7 Body planes. (https://www.shutterstock.com/image-vector/ anatomical-planes-section-showing-sagittal-coronal-314480474.)

ANATOMY

Vertebral Column

Learning the bones that make up the spinal column is important in radiotherapy. For example in radiotherapy imaging as the vertebrae are often used as anatomical landmarks. For example, the carina (a stable piece of cartilage that separates the left and right main bronchus) sits at the vertebral level of T4/5 and is often referred to when matching anatomy on a radiotherapy image.

The vertebral column is split into five parts. The cervical spine, thoracic spine, lumbar spine, sacrum and coccyx.

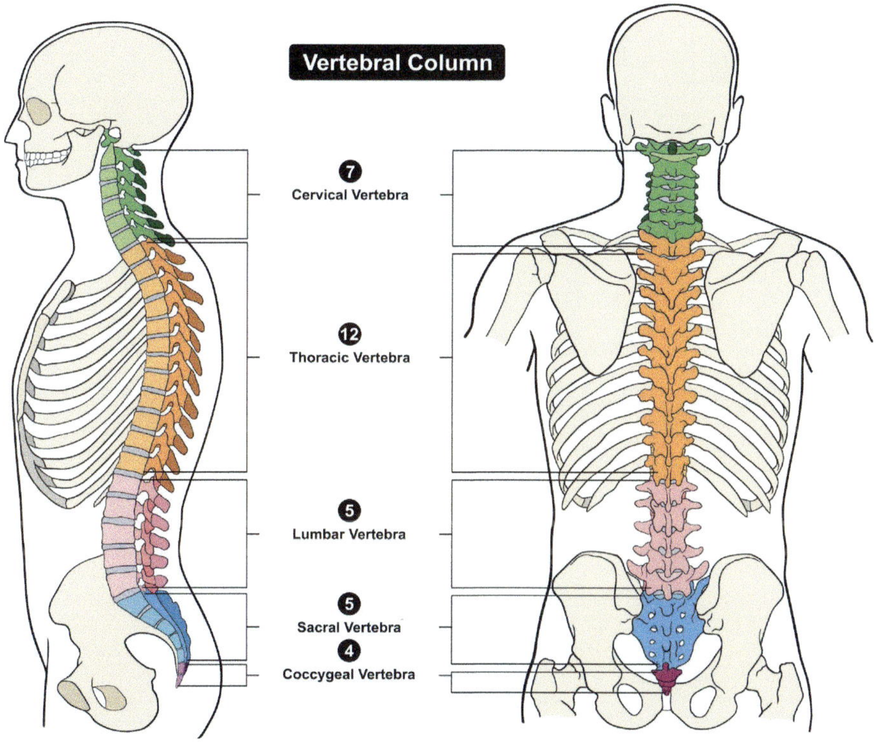

FIGURE 2.8 The bones of vertebral column. (https://www.shutterstock.com/image-vector/vertebral-column-human-body-anatomy-infograpic-703325860.)

Body Cavities

The body has four cavities which contain the bodies organs. These are called the cranial, thoracic, abdominal and pelvic cavity.

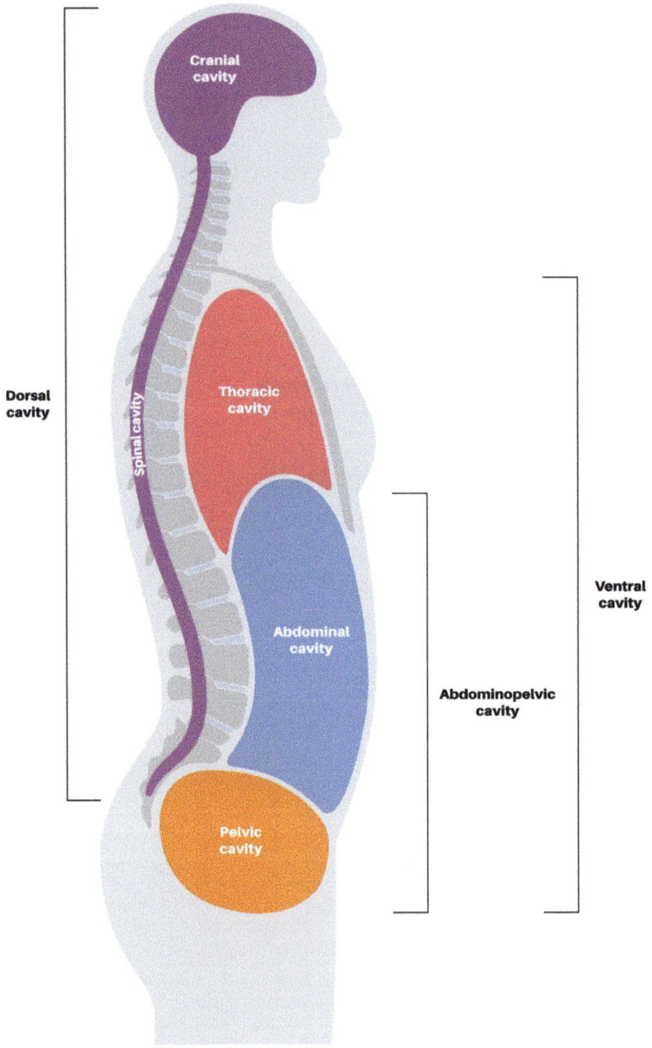

FIGURE 2.9 Body cavities. (Body Cavities Anatomical Poster Spaces Male Stock Vector [Royalty Free] 2306911967 | Shutterstock.)

The **cranial cavity** contains the brain, which is protected by the bones of the skull.

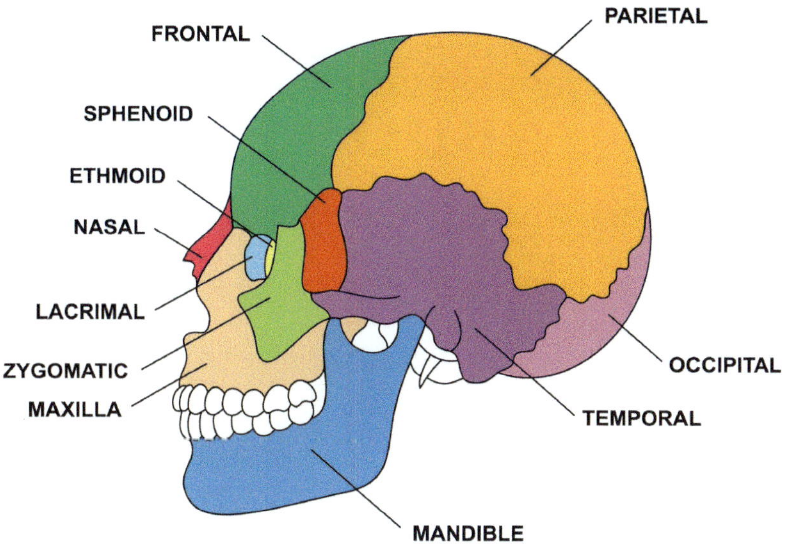

FIGURE 2.10 Cranial bones. (https://www.shutterstock.com/image-vector/human-skull-bones-anatomy-descriptions-colored-2175191535.)

The **thoracic cavity** is the second largest cavity, and the organs are protected by the rib cage. Organs and anatomical structures in the thoracic cavity include the lungs, heart, aorta and oesophagus.

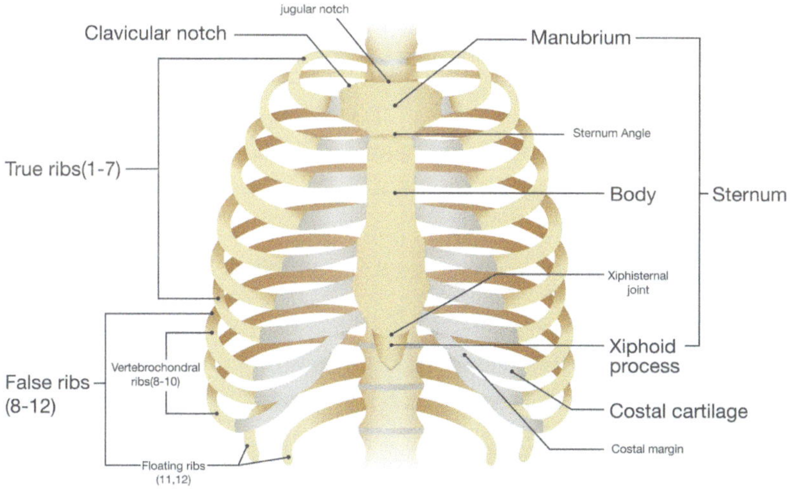

FIGURE 2.11 Rib cage anatomy. (Anatomy Human Rib Illustration Vector About Stock Vector [Royalty Free] 600282818 | Shutterstock.)

The **abdominal cavity** is the largest cavity and contains organs such as the stomach, spleen, pancreas, appendix, kidneys and colon.

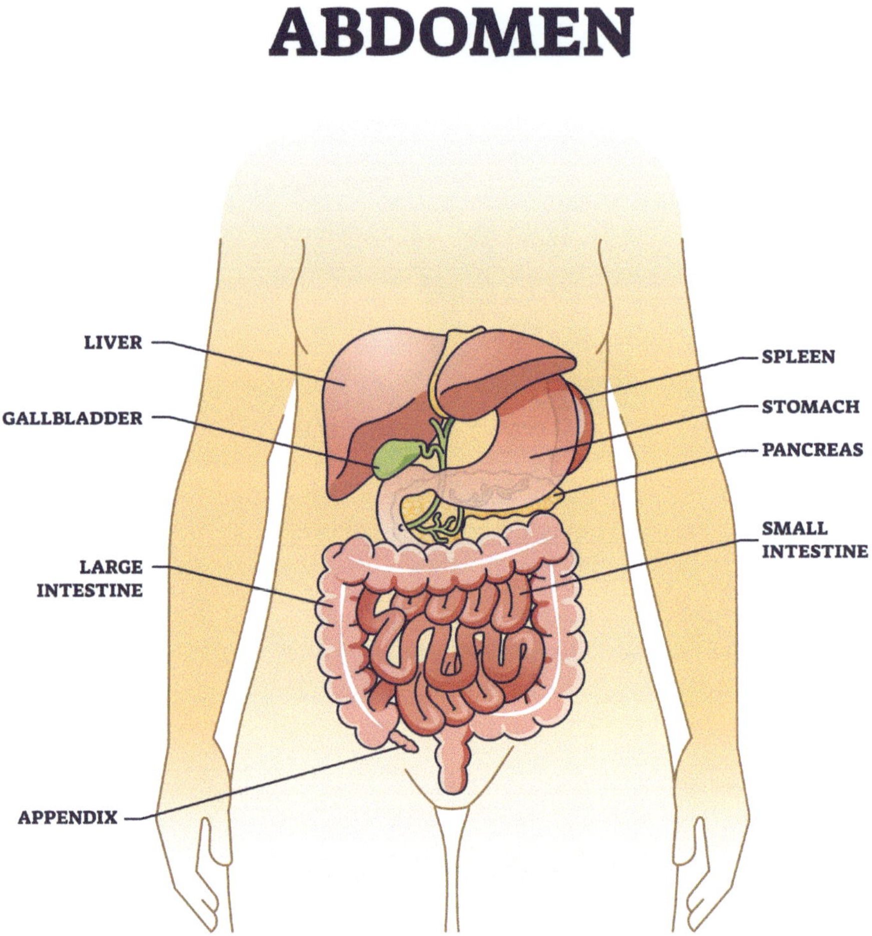

FIGURE 2.12 The organs of the abdomen. (Abdomen Anatomical Structure Inner Body Stomach Stock Vector [Royalty Free] 2385439311 | Shutterstock.)

The **pelvic cavity** differs between males and females and contains the majority of the reproductive organs. The bones of pelvic girdle surround the pelvic cavity.

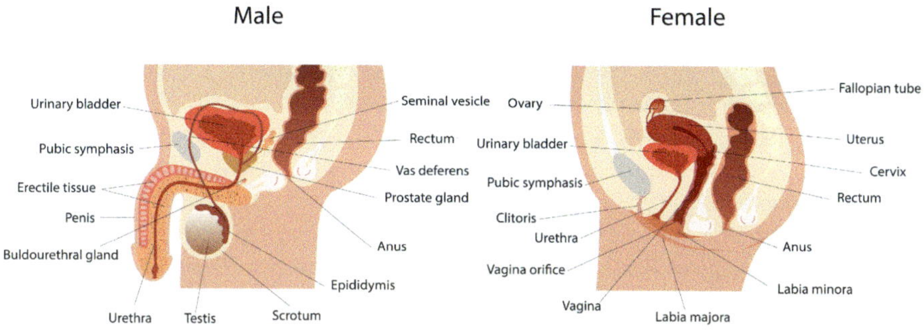

FIGURE 2.13 Organs of the male and female pelvis. (Male Female Reproductive System Median Section Stock Vector [Royalty Free] 1682185156 | Shutterstock.)

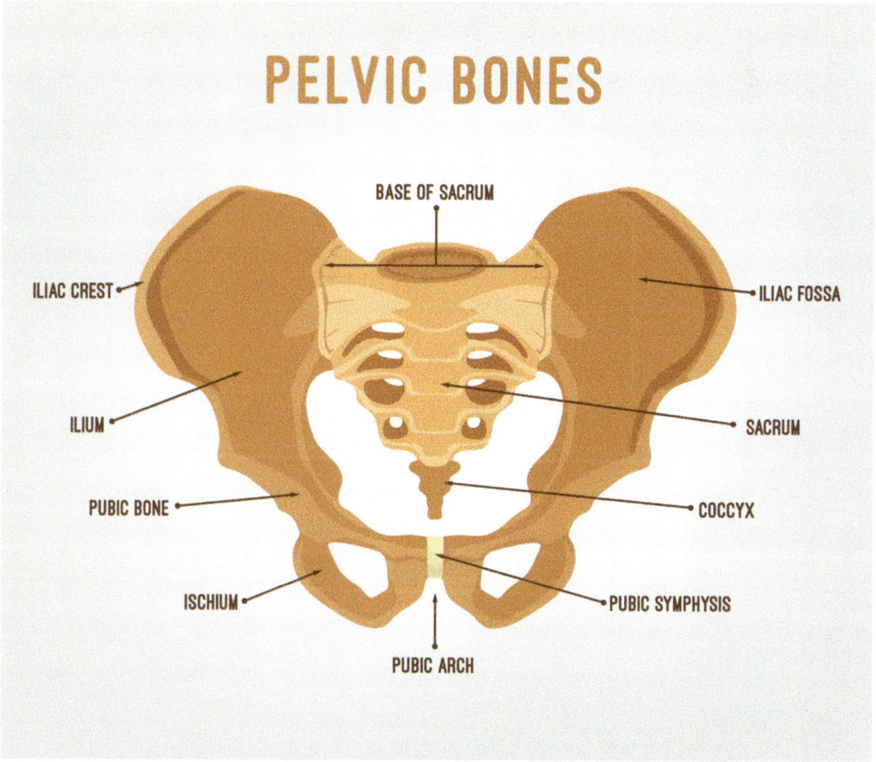

FIGURE 2.14 Bones of the pelvis. (Human Male Anatomy Scheme Main Pelvic Stock Vector [Royalty Free] 772574626 | Shutterstock.)

REFERENCES

Legislation.gov.uk. 2017. The ionising radiations regulations. Legislation. https://www.legislation.gov.uk/uksi/2017/1075/contents (accessed 17 December 2025)

The Royal College of Radiologists (RCR) 2019. The timely delivery of radical radiotherapy: guidelines for the management of unscheduled treatment interruptions. Timely delivery of radical radiotherapy: Guidelines for the management of unscheduled treatment interruptions, Fourth edition | The Royal College of Radiologists (accessed 17 December 2025)

Yonekura, Y., Mattsson, S., Flux, G., Bolch, WE., Dauer, LT., Fisher, DR. *et al.* 2019. ICRP Publication 140: Radiological Protection in Therapy with Radiopharmaceuticals. *Annals of the ICRP 48*, no. 1 (September): 5–95. https://doi.org/10.1177/0146645319838665

CHAPTER 3

Oncological Management

INTRODUCTION TO ONCOLOGICAL MANAGEMENT

Oncological management can consist of either one or many treatments that a patient can receive, with the aim to either relieve symptoms caused by disease or with an aim to cure. In essence, it is how cancer and its symptoms are eradicated or controlled.

This chapter will cover the general principles of oncological management, including what influences management options, common terminology and types of oncological management. Side effects of treatments will also be explored, in particular those of radiotherapy and their management. By the end of this chapter, you will be able to identify the different management options available to a patient and their role in both radical and palliative treatments.

To fully grasp why specific management options are recommended for certain individuals and types of cancer, it is essential to understand how and why cancer develops and behaves.

EPIDEMIOLOGY

Epidemiology is the study of how diseases naturally occur in various populations and why. Rather than focusing on individual cases, these studies examine the overall health of specific groups. This 'natural experiment' involves a three-step process, each providing unique insights to help

DOI: 10.1201/9781003639183-3

scientists build a comprehensive understanding. These steps are known as descriptive, analytical and experimental epidemiology.

Descriptive epidemiology is the first step. It focuses on three factors: person, place and time.

Person-led factors encompass the groups personal attributes such as their age and education. Place refers to geographical placement, and time looks at when and for how long a disease has occurred. This includes illnesses that occur cyclically such as a winter cold and pandemics such as COVID-19.

The second step of epidemiological research is analytical research, which aims to understand why a disease is more common in a particular group. It focuses on the relationship between exposure to specific factors and the prevalence and types of diseases. Unlike descriptive epidemiology, which looks at populations as a whole, analytical research considers individual characteristics, behaviours and habits. The exposures examined in this step include behavioural factors (such as smoking), environmental pollutants (such as asbestos) and genetic traits.

The third and final step is experimental epidemiology. After analysing the results from the first two steps, scientists draw conclusions about what, where and why certain groups of people develop specific diseases. To validate these conclusions, they can monitor individuals who are exposed to a particular factor and compare them to those who are not. For instance, researchers might track smokers and non-smokers to observe whether they develop lung cancer.

The results obtained from epidemiological research are crucial for cancer prevention, as they identify which groups of people are more susceptible to the disease and the reasons behind their increased risk. In other words, epidemiological research highlights the aetiological factors of cancer.

AETIOLOGY

Aetiology is the examination of all aspects involved in disease development, including a person's vulnerability and how cancer-causing agents affect the body. In essence aetiology describes the causative factors of cancer.

Causative factors of cancer can be categorised as follows:

- Environmental: Can be man- made or naturally occurring cancer-causing agents (carcinogens) that are present in the air we breathe and the food we eat, amongst other things. For example, second-hand tobacco smoke.

- Occupational: Carcinogens that are present in certain jobs. For example, a carpenter surrounded by wood dust which can cause cancers in the nasal cavity.
- Viral: Some viruses can cause cancer. For example, the human papilloma virus (HPV) can lead to cervical, anal, oropharyngeal, vaginal and penile cancers.
- Hereditary: Certain cancer types can be passed down through family members due to certain gene faults. For example, BRCA (Breast Cancer gene) 1 and 2 can lead to a greater risk of developing breast and ovarian cancer.
- Physiological: Aspects of a person's lifestyle, for example, eating and exercise habits and body mechanisms, for example, hormones and inflammation.

Another aetiological factor to consider is referred to as iatrogenic. This is when a medical intervention can lead to disease. For example, the use of radiation to cure cancer can also lead to cancer. Aetiology plays a key role in cancer prevention and early detection through screening. Epidemiological research has established that specific factors, environments or personal characteristics can contribute to cancer development. With this understanding, individuals can make informed lifestyle choices and choose to undergo screenings or preventive treatments to reduce their risk. However, it is not always possible to avoid all carcinogenic agents.

SIGNS AND SYMPTOMS

Signs and symptoms often indicate that changes have occurred in the body. They arise when cells undergo alterations, and if a significant number of cells are affected, these changes can impact the function of an organ. Symptoms are subjective experiences that cannot be seen, such as a headache or fatigue. In contrast, signs are objective and observable, like a lump or an abnormal blood test. However, in some cases, no signs or symptoms are present. This is known as being asymptomatic. This is why screenings are crucial, as many cancers and other diseases can remain undetected without noticeable symptoms until they have grown significantly or spread to other areas of the body.

SCREENING

Screening is a public health service that aims to identify diseases and abnormalities in at risk populations, who are generally in good overall health. If a disease is detected early, there is a higher chance that it can be treated more effectively.

Population, targeted and stratified are the three types of screening. Population screening is provided to a specific group of individuals selected from the general population based on demographic factors such as age and sex. For example, cervical screening in the UK is offered to anyone with a cervix between the ages of 25–64.

Targeted screening is a proactive approach aimed at individuals or groups with a higher risk of developing certain conditions due to their lifestyle choices or existing health issues. An example of this is the NHS lung health check program. Lastly, stratified screening: this approach tailors the frequency and type of screening tests based on an individual's level of risk. It can be used to enhance both targeted and population screening programs. An example of stratified screening is the annual diagnostic screening of people with breast tissue who have been identified as having the BRCA 1 and 2 genes.

The type of investigations a person may undergo depends not only on their signs and symptoms but also on the results of the screening.

INVESTIGATIONS

There are many kinds of investigations a person can be offered. The first investigations a person is usually offered will take place upon presenting with a sign and or a symptom at their general practitioners (GP). These include palpation (examination through touch), percussion (examination through tapping body parts), auscultation (listening to body sounds) and vital signs (these include blood pressure and temperature).

Diagnostic investigations are a secondary type of examination that individuals may undergo to assess the stage, grade and location of cancer. To gain a comprehensive understanding of the cancer and its characteristics, patients are often offered a range of diagnostic tests. Each test provides unique insights. Biochemistry provides data on the body's biological mechanisms via blood tests; cytology explores how and why cells have changed and imaging and endoscopy pinpoint where the cancer is and if it has spread. Imaging modalities include, but are not limited to, computed tomography (CT), magnetic resonance imaging (MRI) and positron emission tomography (PET).

METHODS OF SPREAD

If a group of abnormal cells has not spread from the location, they have formed, in this is referred to as carcinoma-in-situ. A cancer can be classed as malignant if it has the right characteristics that indicate it could spread or has spread to other parts of the body. The spread of cancer can be further classified as local or distant. Distant spread means that the cancer has metastasised from one body part to another that is far away from the organ of origin. Cancer that invades nearby structures is known as local or direct spread.

Commonly, a cancer will spread locally. This means that the cancer continues to grow at its origin.

Tumour cells can metastasise around the body via the lymphatic system; cancer cells can invade the intricate lymph system to travel to nearby or faraway organs. The main function of the lymphatic system is to move lymph back into the bloodstream and protect the body from infection. Another intricate body system cancer cells can use to travel around is the circulatory system. Specialised cells called circulating tumour cells enter the thin walls of capillaries, and they can then circulate around the body and travel through the vessel into an organ. A less common but clinically significant route of cancer spread is transcoelomic spread. This occurs when cancer cells travel across body cavities.

Cancer cells can also be implanted during surgery and investigations via a scalpel or needle. When an instrument is used multiple times to pick up a sample, it can pick up cancer cells and move them. This is called seeding. It is important to remember that methods of spread vary due to cancer type and its position in the body. It is also common for cancers to utilise one or more methods. Tumour spread, or capacity to spread, greatly influences how their stage and grade are determined.

STAGING AND GRADING

Staging and grading are very important disease-led factors that help determine the type of oncological management a patient may receive. Staging encompasses the tumours cell type, size, spread and location. There are many different types of staging, and some are unique to certain cancer types. The most common staging system is the Tumour (T) Node (N) Metastases (M) system. T stands for the size and extent of the disease,

N stands for how many lymphatic nodes have been invaded, and M represents any distant spread.

Grading is based on the microscopic appearance of the cancer cells. Generally, cancers are graded from 1 to 4; however, some cancers can be graded at a 5. The lower the grade the more, 'normal' or well-differentiated the cancer cells appear. The higher the number, the more abnormal or poorly differentiated they appear.

DISEASE-LED AND PATIENT-LED FACTORS

The type of oncological management a patient can receive depends on many different patient-led and disease-led factors. Disease-led factors encompass the size, location, type and stage and grade of the tumour. Patient-led factors encompass any co-morbidities the patient may have, their beliefs and their lifestyle. Patient choice also plays a huge role in the type of oncological management they will receive. Ultimately, patient choice is the deciding factor for any type of medical intervention. A patient must consent to a procedure and can remove their consent at any time.

One of the most important elements that is considered is how the treatment will affect the patient's quality of life (QOL). QOL encompasses how comfortable, healthy and fulfilled a patient's life will be post (after) treatment. These factors determine whether someone is likely to receive palliative or radical treatment.

PALLIATIVE AND RADICAL TREATMENT

Treatment intent falls into two categories; palliative and radical. Radical describes any medical intervention given with an aim to cure. In other words, it results in the patient becoming cancer free or in remission. Radical treatments usually involve more complicated and intense interventions than palliative treatment.

Palliative treatment describes any medical intervention given with an aim to improve a person's remaining QOL. Unfortunately, it is not always possible to 'cure' cancer, due to various patient-led and disease-led factors. Instead of aiming to cure, palliative treatments aim to improve symptoms of cancer such as pain or bleeding. Patients under palliative care may have weeks, months or years to live.

It is important to remember that both radical and palliative treatments are equally important. Although they may have different aims, they both have the potential to improve someone's QOL. Both radical and palliative regimens can involve adjuvant, neo-adjuvant and concomitant oncological management options.

PRIMARY, ADJUVANT, NEO-ADJUVANT AND CONCOMITANT TREATMENTS

A primary treatment can be used by itself, before another management option such as surgery, or can be used in conjunction with another treatment such as chemotherapy or radiotherapy. Primary treatment is always the main treatment.

Neo-adjuvant treatment is any treatment given before the primary treatment. Often this is done with the aim to shrink or reduce the tumour, such as a de-bulking surgery.

Adjuvant treatment refers to any treatment given after the primary treatment. The aim of adjuvant treatment is to reduce the risk of the cancer returning. Hormone therapy is a good example of this. Sometimes a primary treatment can become more effective if it is coupled with another treatment. For example, some cancer types are not radiosensitive. To make them more susceptible to radiation, chemotherapy is often given alongside radiotherapy. This is a type of concomitant treatment.

TREATMENT OPTIONS

Surgery

Surgery is often utilised for a variety of reasons and is often a first-line management option, meaning that it will be the first and possibly the primary treatment. As part of an investigation, surgery can be used in the form of a biopsy. Surgery is also used for both palliative and radical management regimens. The main aim of surgery is to provide local and regional control. However, it can often be used palliatively to help relieve symptoms of a large or invasive tumour.

One common type of surgery a patient will receive is called de-bulking surgery. De-bulking means that the surgeon aims to remove as much of the tumour as possible. This may result in a part, a whole or several organs being removed. Surgery can also be used prophylactically. Prophylactic surgery is also referred to as risk-reducing and preventative surgery. To do something prophylactically is to be pre-emptive. Certain biological markers, such as the BRCA 1 and BRCA 2 gene variants, are well known to increase the risk of developing breast cancer. To decrease this risk, a prophylactic double mastectomy can be offered to the person carrying that gene.

Chemotherapy

Chemotherapy is a systemic therapy that uses one or more cytotoxic drugs to target the cell cycle. Unlike normal, healthy cells, cancer cell growth can be unregulated and can replicate abnormally quickly. Some cytotoxic drugs are designed to target a specific phase of the cell cycle to disrupt this replication.

Some chemotherapy regimens will involve a mix of different drugs, as different drugs target and disrupt different phases of the cell cycle. Due to the nature of cytotoxic drugs, all radiotherapy departments have access to a spill kit to ensure patient and staff safety.

The type and length of the chemotherapy prescription vary from person to person due to disease-led and patient-led factors. Some patients are not able to receive certain drugs due to existing co-morbidities such as kidney failure. If this happens, they are often able to receive a different cytotoxic drug.

Chemotherapy can be given in a variety of ways; these are called routes of administration. The route of administration depends on three main factors: cancer type, location of cancer and the type of chemotherapy being administered.

Routes of Administration

Most common:

- Oral: Pill form
- Topical: A cream that the patient rubs onto their skin
- Intra-arterial (IA): Injected into an artery
- Intravenous (IV): Into the bloodstream via an injection, drip or infusion pump

Chemotherapy can be administered in different ways. Some administration methods can also double up to allow a health care professional to draw blood and administer different drugs such as anti-emetics and pain relief:

- Cannulas: Small tubes inserted into the body
- Hickman (central) lines: Tube inserted into the jugular vein
- Peripherally inserted central catheter (PICC): Made up of 2–3 small tubes that join together as one tube inside the body. This allows the patient to receive different treatments at the same time
- Portacath: Small tube inserted into the superior vena cava and connected to a port under the skin

Less common:

- Intravesical: Into the bladder
- Subcutaneous: Injections under the skin
- Intramuscular: Injections into the muscle
- Intrathecal: Injected into the spine
- Intrapleural: Injected into a pleura
- Intraperitoneal: Injected into the peritoneum

Biological Therapy

Biological therapies utilise substances from living organisms to treat a variety of diseases, including cancer. These substances can be man-made (synthetic) or occur organically in a human body. There are two main classifications of biological therapy: targeted therapy and immunotherapy. Targeted therapy is made up of very small molecules. These molecules attack the cancer cells directly by interacting with molecular targets involved in cancer development and progression.

In contrast, immunotherapy does not directly attack cancer cells. Instead, immunotherapy works in conjunction with the body's immune system to make it strong enough to injure and kill the cancer cells. Certain drugs can be classed as both immunotherapy and targeted therapy. An example of this would be monoclonal antibodies (mABs). They can trigger the immune system to attack, and they can also interfere with the proteins that cancers need to grow. Certain mABs can carry drugs or radioactive molecules straight to the cancer cells to kill them.

Radiotherapy

As previously mentioned, radiotherapy is the use of ionising radiation to kill or slow down the production of cancer cells, whilst sparing healthy tissues that surround the tumour or tumour bed. Each treatment is tailored to the patient to ensure gold-standard care. Radiotherapy can be given in different forms. In the introduction, it was noted that brachytherapy is a type of internal radiation. This treatment can be used alone, or as a boost, as seen below in the example prescriptions given. Brachytherapy can also be called short therapy, due to its short fractionation schedule, and can be further broken down into: low dose rate (LDR) and high dose rate (HDR). These terms refer to the rate at which the radiation is delivered, which depends on the radiation source used. Both LDR and HDR use radioactive sources that are placed inside the patient's body. However, LDR sources are permanent sources in the form of small seeds that emit doses of radiation over time and will not be removed. HDR brachytherapy is temporary. This type of brachytherapy is inserted via a system of tubes and or needles and is removed after the treatment.

RADIOTHERAPY PRESCRIPTIONS

Like most medical treatments, radiotherapy uses prescriptions to ensure treatment efficiency and limit side effects. Radiotherapy prescriptions are carefully selected and vary depending on disease type and location. Most tumour types have a 'standard' prescription that has been established through research and evidence-based practice to achieve biological effectiveness with the least side effects. However, these can vary depending on disease and patient-led factors. New prescriptions are continuously being trialled to ensure that patients always receive the best treatment for them.

In the UK, the Royal College of Radiologists (RCR) has produced guidelines on prescriptions. The first, 'Radiotherapy dose fractionation', guidance was published in 2006 to help standardise radiotherapy prescriptions and is updated regularly, including most recently the 4th edition in 2024 (RCR 2024). Typically, a patient receiving radical EBRT will receive a higher prescription than a patient receiving palliative EBRT. Similarly, due to the differences in treatment type, a patient undergoing brachytherapy will receive less fractions than a patient undergoing EBRT. It is important to note that prescriptions may differ from patient to patient even if they have the same disease. Some patients may also receive multiple prescriptions, such as a radical prostate patient (see example below).

TABLE 3.1 Example Prescriptions Taken from RCR Dose and Fractionation Guidelines 4th Edition 2024 for a Patient Receiving Treatment to Their Prostate Only

Prescription	Dose in Gy	Fractions	Time Frame	Brachytherapy
Radical EBRT	60	20	4 weeks	
Radical EBRT+boost	37.5	15	3 weeks	15Gy HDR
Palliative	20	5	1 week	

Source: RCR (2024).

After the prescription has been decided upon, a patient will be invited to a pre-treatment appointment.

PRE-TREATMENT

Pre-treatment is the first practical step of a radiotherapy treatment plan and therefore can also be referred to as the planning appointment. During this appointment, radiographers will go through the patients' consent form that was previously given to them by their oncologist. This will include the patients' personal details, such as their date of birth and pregnancy status. It also includes the anatomical treatment area, the process of radiotherapy and the potential side effects. In addition, this is usually the first opportunity a patient has to practice any pre-treatment preparation.

Pre-treatment Preparation

Some patients require pre-treatment preparation, for example, it is an essential step for most patients undergoing pelvic radiotherapy. These patients will be asked to either have an empty or full bladder and empty rectum. This can be achieved through the use of drinking protocols and enemas. Pelvic anatomy such as the bladder and rectum can differ greatly in size day to day, depending on several factors including diet. The size of this anatomy can affect the movement of the treatment volume. For example, in the treatment of a radical prostate patient, a full bladder helps to push the small bowel up and out of the pelvis, so it is moved away from the radiotherapy treatment area. An empty bowel means the other structures near and around the rectum are less likely to move. Not only does this help improve treatment accuracy, but it also helps to decrease certain side effects. There is no set guidance on how much a patient should drink and departmental protocols can vary.

Once consent is obtained, the patient will be asked to lie down on the CT couch, and the radiographers will select the correct immobilisation for them. If the patient is receiving EBRT, they will then apply radio-opaque markers that can be seen on the CT scan to simulate where the permanent marks will be placed. This is part of the localisation process.

IMMOBILISATION

Immobilisation is the term used to describe the equipment used that helps to keep a patient still, straight and comfortable. The type of immobilisation used is determined by the location of the tumour, the patient's anatomy and to some extent the type of radiotherapy technique being used. The same immobilisation will be used for each fraction to help maintain a reproducible treatment position. Some immobilisation can also be personalised to fit the patient. This piece of equipment will only be used for that patient, for example a head and neck shell. Please note that the names of the immobilisation shown below may be different in your clinical department; however, they fulfil the same purpose.

Examples of Immobilisation

Breast board, knee rest and head rest: This enables the patient to comfortably lift their arms up and away from the treatment area (the breast). The wedge angle can also be adjusted to allow for the beam angle. At the bottom of the breast board, you can see a bump; this is called a butt stop and is there to stop the patient from sliding down the bed. A knee rest is also used for comfort and to help the patient's back lie flat against the board. A knee rest is one of the most used pieces of immobilisation and can be used for most treatment sites.

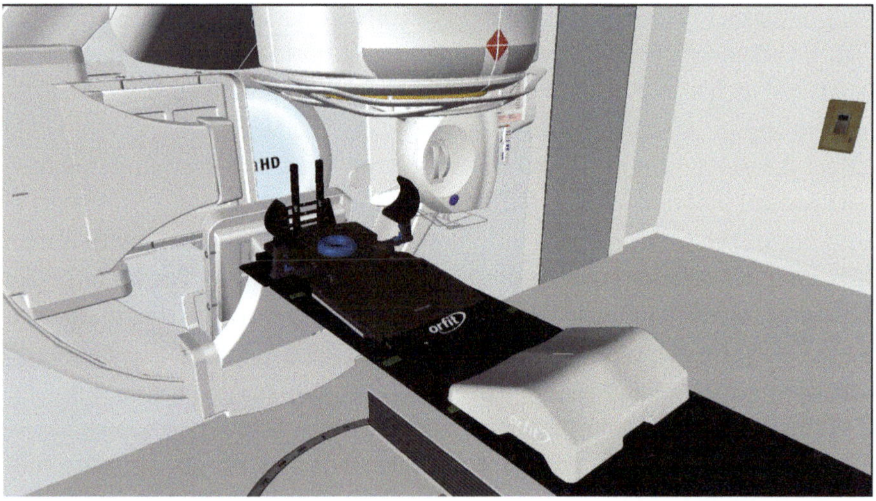

FIGURE 3.1 A breast board, knee rest and head rest. (With permission from Vertual.)

Wing board with head rest: This is a versatile piece of equipment and can be used to treat lung, lymphoma and some gynaecological patients who have a long treatment field. It can also be used to treat lower third oesophageal patients. Similar to a breast treatment, the patient is required to move their arms up and away from the treatment site. A head rest is used to ensure the patient's head and neck is supported.

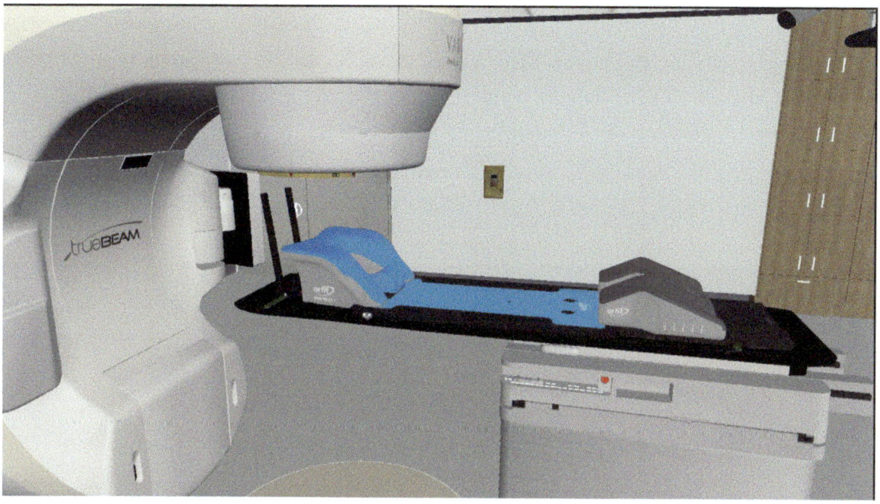

FIGURE 3.2 A wing board. (With permission from Vertual.)

Combifix with knee rest and ankle rest: A Combifix (bottom panel) is used as part of an indexing system for a knee and ankle rest. A knee rest can be lifted up using small cushions, d the angle of an ankle rest can also be changed. This is commonly used for any patient receiving treatment to their pelvic area. A head rest will also be used.

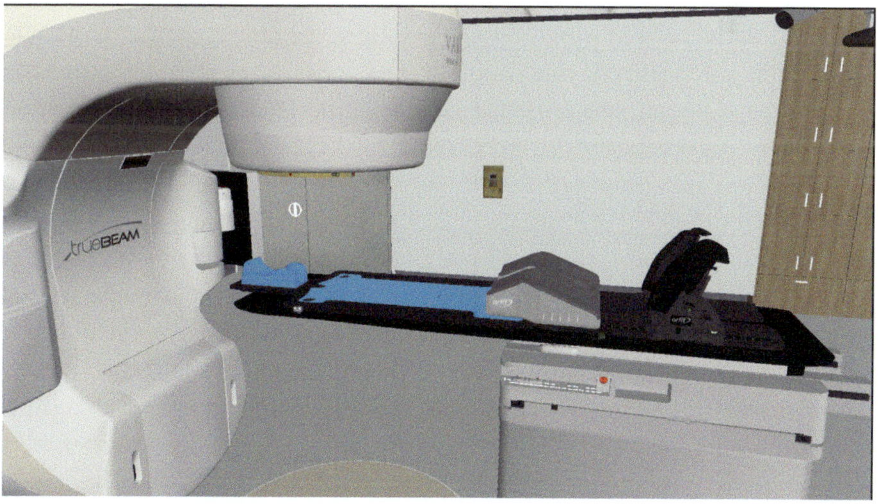

FIGURE 3.3 A Combifix. (With permission from Vertual.)

Head and neck shell: This is a personalised piece of equipment and is moulded to the patient's face and neck. A shorter version is also used (without the neck) for brain patients. As there are many small structures in the head, referred to as organs at risk (OAR), it is extremely important that the patient stays still. A shell combined with a head rest ensures the patient cannot turn or nod their head.

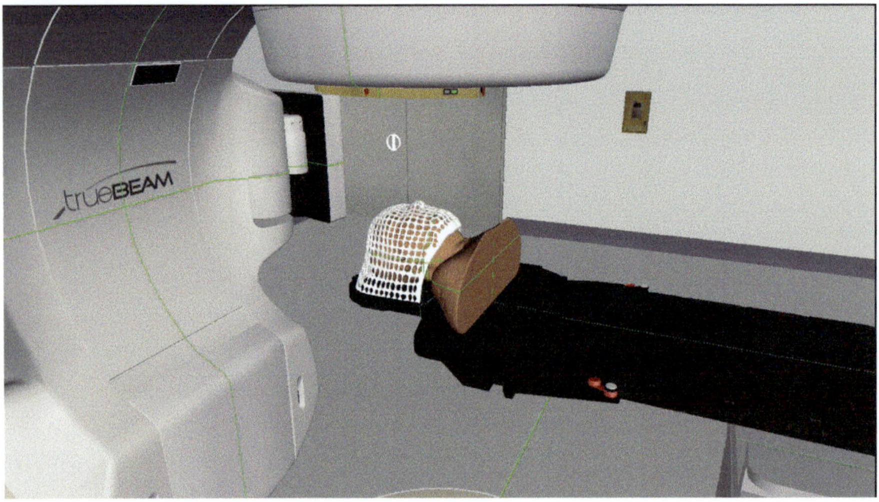

FIGURE 3.4 A head and neck shell. (With permission from Vertual.)

Once the correct immobilisation is chosen a scout scan/topogram checks that the patient is in the right position, the radiographers can make adjustments if required. If no adjustments are needed, they will continue to take the CT scan. Once the CT scan is completed, they will then apply permanent marks (tattoos) if necessary.

PERMANENT MARKS

Permanent marks, also referred to as tattoos, are placed on a stable area or areas of the patient. Please note: permanent marks are an extremely important part of the pre-treatment process and will be further explained in Chapter 6. Usually, a patient will receive three, one anterior and two laterally. These marks will be placed near to the treatment area. However, this is not where the treatment will enter the patient. Instead, these marks are used to make sure the patient is straight, and they are in the same position for each treatment. A small needle and ink are used to apply the mark; the smaller the mark the better, both for precision and cosmetic effect. Some clinical sites do not use permanent marks or tattoos. These departments use surface guided radiotherapy (SGRT) which scans the patient's body with thermal cameras to ensure they are in the right place by mapping out the patient's skin surface. Once the pre-treatment appointment is complete, the scans and patients' details are sent over to planning where a treatment plan is made.

EXTERNAL BEAM TREATMENT TECHNIQUES

All treatment techniques aim to target the treatment, giving a homogenous dose to the treatment site, whilst minimising radiation exposure to OAR. OARs are healthy organs that surround the planning target volume (PTV). Each OAR can receive a certain amount of dose before side effects or permanent damage occur. To reduce radiation exposure, certain techniques are used for certain sites in correlation with beam dose and energy, as well as the use of shielding.

BOLUS AND BUILD-UP

Bolus is made from different materials, such as Superflab. It has a set jelly-like texture and typically comes in sheets that can be cut to size if required. It can be used in any circumstance where treatment needs to be close to the skin surface, such as treating lymph nodes. In order to achieve this, bolus acts as a tissue substitute enabling the dose to be deposited closer to the skin surface.

Build-up is made of different materials, usually a waxy substance. The goal of build-up is to fill in any gaps or crevices in the body, such as the gap behind the ear, to achieve a more homogenous dose.

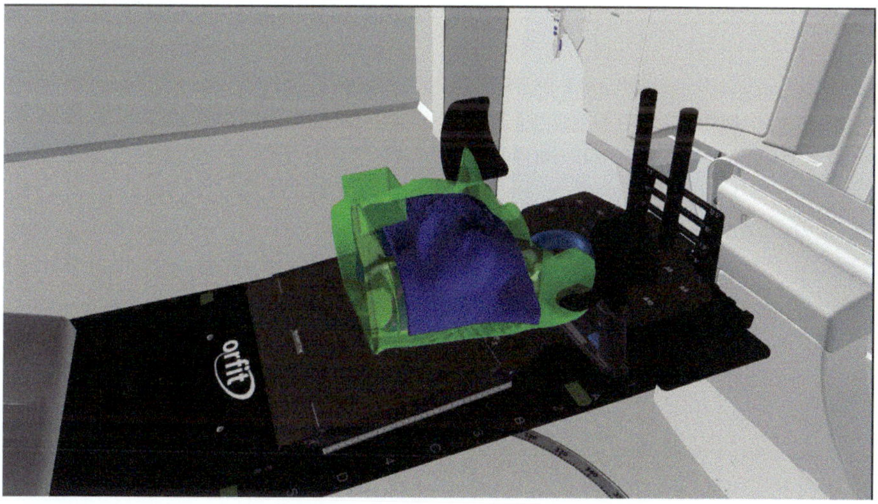

FIGURE 3.5 A bolus on a patient. (With permission from Vertual.)

RADIOTHERAPY SIDE EFFECTS

A side effect is classed as an unintended effect of a medical treatment. Radiotherapy aims to damage cancer cells; unfortunately, healthy cells that surround the treatment area can also be affected. Rapidly dividing cells are most susceptible to the side effects of radiation. These include skin cells, cells lining the gastrointestinal tract and the lining of the mouth (oral mucosa).

There are two classifications of radiation induced side effects: acute and chronic. Acute side effects may occur during treatment or up to three months post treatment. They are usually reversible and lessen in intensity over time. Late or chronic side effects are long-lasting and can be irreversible. They can occur months or even years post-treatment and have the tendency to get worse over time.

The type and severity of side effects a patient may experience depend on various factors, including the treatment modality, tumour location and the proximity of surrounding OARs. For instance, a patient undergoing radical EBRT to the prostate may develop changes in bowel and bladder function due to the close proximity of these organs to the PTV. In contrast, a patient receiving radiotherapy to the oral cavity is likely to experience different side effects, such as mucositis, due to the specific anatomical structures in that region. In this chapter, we will be exploring the two most common acute side effects of radical radiotherapy.

It is important to note that not all patients will experience the following side effects. The type and severity of side effects differ from person to person due to a number of patient and disease-related factors. All patients will have access to many members of the multidisciplinary team (MDT) throughout their treatment, such as a specialist nurse, radiographers and their oncologist, who can help them manage any adverse effects of their treatment. If a patient displays any side effects, a therapeutic radiographer will always communicate this to the rest of their team and document in the patient's notes to ensure the patient receives the best care. A therapeutic radiographer can give some advice on side effect management; however, if it is out of their scope of practice, they will then refer the patient onto someone within the MDT who is better placed to give them support, this is called signposting.

The most common radiotherapy side effect, across all sites, is fatigue. Fatigue is a state of extreme tiredness which results in highly decreased energy levels and body weakness. Fatigue is usually an acute side effect; however, in some cases, it can develop into a chronic side effect. Advice around alleviating fatigue focuses on light exercise, resting when needed, eating regularly and staying hydrated. A patient may also be referred for a blood test to check that they are not anaemic. Complimentary therapies such as massage and aromatherapy can also help some patients manage their fatigue.

Erythema is the second most common side effect of radiotherapy. Erythema is a name used to describe changes in the skins appearance and texture. It is also referred to as radiation dermatitis. There are different stages of erythema, and they are assessed and managed via a tool provided by the Radiation Therapy Oncology Group called the RTOG system. Table 3.2 has been adapted from the Society and College of Radiographer's guidelines (SCoR 2020).

TABLE 3.2 RTOG System

Grade	Grade 0	Grade 1	Grade 2a	Grade 2b	Grade 3
Definition	No visible changes to skins surface	Faint or dull erythema	Tender or bright erythema	Patchy moist desquamation	Confluent moist desquamation
Description	Skin appears normal	Slight redness may develop, along with a feeling of tightness or mild itching	The skin may appear redder and feel tight and sensitive, itchy and/or sore	Some areas of skin begin to break down, potentially showing yellow or pale green fluid. Swelling and oedema are present	Larger, connected areas of skin are broken down with clear signs of yellow or pale green fluid. Swelling and oedema are also clearly noticeable

Source: SCoR (2020).

All patients undergoing radiotherapy will be encouraged to practice self-care to keep their skin healthy. Self-care is as follows: protect skin from the sun, moisturise treatment area daily, wear loose non-itchy material such as cotton. Shower and bathe in lukewarm water and pat-dry. Do not add extreme temperature such as ice packs onto treatment area.

If a patient is experiencing itchiness and pain in the treatment area, they can be prescribed steroid cream to help combat the itchiness and analgesia for the pain. If the skin starts to break down, a swab will be taken to determine whether an infection is present. Patients will also be advised to keep the broken area clean and dry. In higher RTOG grades, a dressing may also be necessary. Once a patient has completed their course of radiotherapy, they then move onto follow up care.

CANCER PATIENT PATHWAY

The following section explores the key stages of the cancer patient pathway, from initial presentation through to post-treatment aftercare. Now that you are familiar with oncological management and the steps involved in diagnosing, staging and treating cancer, you can better appreciate how each stage influences the patient's journey.

This understanding is essential so that you can understand the pathway that the patients you see in radiotherapy have taken, helping you to give them compassion and appropriate support. This section also outlines radiotherapy-specific processes and highlights National Health Service (NHS) and the National Institute for Health and Care Excellence (NICE) referral guidelines designed to support timely and effective treatment delivery.

REFERRAL GUIDELINES

Referral guidelines ensure that everyone receives their diagnosis and subsequent treatment in a timely manner and can be seen below.

- The 28-day Faster Diagnosis Standard (FDS)
- 62-day referral to treatment standard (RTT)
- 31-day decision to treat to treatment standard (DTT)

There are three referral brackets a patient can go through as defined by the NICE Rapid Referral guidelines (NICE 2025). These brackets are:

- Non-urgent: Person requires routine referral or tests
- Urgent: Two-week wait urgent referral pathway
- Very urgent: Referral required within 48 hours

The cancer patient pathway adheres to a strict structure to ensure prompt and effective care that is tailored to each individual's needs. Below is a summary of the key stages involved in this process. It is important to note that the patient and their family/carers are involved and informed at each step.

1. **GP visit:** This can be due to a patient developing certain signs or symptoms or for a routine screening. The GP will assess these signs and symptoms, as well as any positive screening results, and refer them onto the appropriate pathway (see 'referral guidelines' for more information).

2. **Diagnostic investigations:** Various cytology, imaging and biochemistry tests will be performed to determine a diagnosis.

3. **Staging and grading and MDT:** If a patient is diagnosed with cancer, the disease stage and grade will be determined and discussed in an MDT meeting. In this meeting, treatment options such as surgery and radiotherapy with either a radical or palliative intent will be discussed. The patient is then informed of the diagnosis and oncological management options available to them. Once the patient has been informed off all options, the positives and negatives of each one and agree that they understand these, consent will be taken.

4. **Treatment planning:** Treatment options will be discussed, including the benefits and drawbacks of each option. Then the best course of treatment is decided upon and planned. As discussed earlier in Chapter 3, patients may receive a number of different treatments.

5. **Pre-habilitation:** Some patients may be offered pre-habilitation once they have received their diagnosis. Pre-habilitation aims to give the patient the tools they may require helping them stay healthy throughout their treatment. For example, diet and exercise advice. It is also very common for patients who smoke to be offered smoking cessation advice and support.

6. **Pre-treatment assessments:** Before certain procedures, such as chemotherapy, routine assessments such as kidney function tests are commonly performed to ensure the patient's body can tolerate the treatment. For patients receiving radiotherapy to the head and neck region, a referral to a specialist dentist is typically offered, as radiotherapy can impact dental health and the integrity of the teeth. These pre-assessments are essential to confirm that the patient is in an optimal state of health to safely undergo the proposed treatment.

7. **Treatment begins**

8. **Follow-up or palliative care:** After-care does not have a definitive timeline and depends on the treatment and the person. Typically, after-care involves a 6-week post-treatment check-up. Patients may also be offered diagnostic scans for up to 5 years. Patients will also always have access to support groups and other types of holistic and complementary care.

RADIOTHERAPY SPECIFIC TREATMENT PATHWAY

This pathway is as above, but with a few changes at steps 4 and 6. These steps are discussed in more detail below. It is important to remember that throughout these processes, ongoing patient care is required. Therapeutic radiographers regularly assess the patient's well-being using effective communication strategies, ensuring any concerns or side effects are promptly addressed.

- Pre-treatment appointment
  The patient will attend a pre-treatment appointment. If required, a separate appointment will be scheduled in the mould room. During this session, specialised immobilisation equipment such as a head and neck shell or electron cut-out will be prepared to ensure accurate treatment delivery.

- Dosimetry planning
  Following the pre-treatment appointment, all relevant clinical and diagnostic information is sent to the dosimetry team. Dosimetrists collaborate with physicists, oncologists and therapeutic radiographers to create an optimal radiotherapy treatment plan tailored to the patient's needs.

- Treatment begins
  On the first day of treatment, a 'first day chat' is conducted. This involves therapeutic radiographers re-confirming patient consent, reviewing the treatment schedule and discussing potential side effects and their management, including appropriate skin care.

REFERENCES

National Institute for Health and Care Excellence (NICE) 2025. Suspected cancer: recognition and referral. Guidelines. Overview | Suspected cancer: Recognition and referral | Guidance | NICE (accessed 19 March 2025)

The Royal College of Radiologists (RCR). 2024. *Radiotherapy dose fractionation*. 4th ed. https://www.rcr.ac.uk/our-services/all-our-publications/clinical-oncology-publications/radiotherapy-dose-fractionation-fourth-edition/ (accessed 18 December 2025)

The Society and College of Radiographers (ScoR). 2020. Practice guidelines. 2nd revised ed. *Radiation dermatitis guidelines for radiotherapy healthcare professionals* | SoR (accessed 23 March 2025)

CHAPTER 4

Physics and Radiotherapy Treatment Planning

INTRODUCTION TO X-RAY PHYSICS

X-rays are a form of electromagnetic radiation with wavelengths much shorter and energies much higher than visible light. Since the discovery of X-rays by Wilhelm Conrad Röntgen in 1895, they have become an essential tool in medicine utilised in diagnosing bone fractures to guiding complex radiotherapy treatments.

Understanding the basic physics behind X-rays and the X-ray tube is fundamental for anyone working in medical imaging or radiotherapy. A clear grasp of how X-rays are produced, how they interact with matter, and how their properties can be controlled allows radiographers to optimise image quality, deliver effective treatments and maintain patient and staff safety.

This section will explore the core principles of X-ray physics, beginning with their physical characteristics and moving through their production, interactions and clinical relevance. While the focus will be on practical application, the scientific foundations will be explained in a way that builds a strong conceptual framework helping you connect the theory to practice.

DOI: 10.1201/9781003639183-4

FUNDAMENTALS OF X-RAYS

As mentioned above, X-rays are a type of electromagnetic radiation, a form of energy that travels in waves at the speed of light. X-rays sit on the electromagnetic spectrum between gamma rays and ultraviolet light. X-rays are distinguished from gamma rays by their origin, gamma rays are produced by nuclear decay from an atomic nucleus and X-rays are produced by interactions with electrons. Unlike visible light, X-rays have a higher energy and can pass through objects, like tissues in the body. Like all forms of electromagnetic radiation, X-rays can be described by their:

1. Wavelength
2. Frequency
3. Energy

X-rays have high frequencies, short wavelengths and high energy compared with visible light. X-rays are a type of ionising radiation, which means when interacting with matter, they are energetic enough to cause neutral atoms to eject electrons. Through this ionisation process, a portion of the X-ray energy is deposited in the matter. When passing through living tissue, X-rays can cause harmful biochemical changes in genes, chromosomes and other cell components. Radiotherapy practice takes advantage of this and uses high-energy X-rays in a beneficial way.

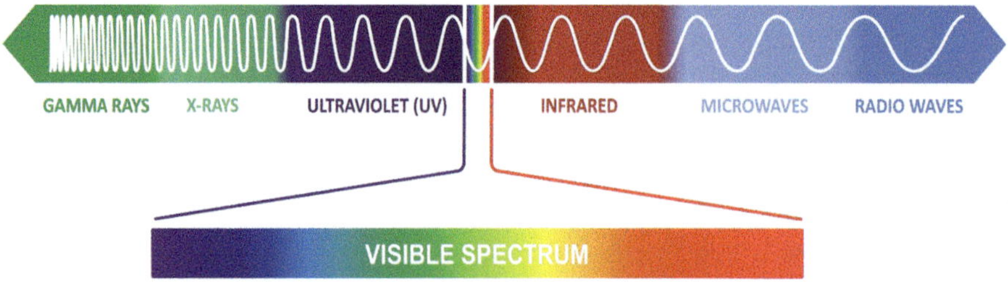

FIGURE 4.1 The electromagnetic spectrum. (Electromagnetic Spectrum Diagram Vector Color Illustration Stock Vector [Royalty Free] 2513673809 | Shutterstock.)

PRODUCTION OF X-RAYS

In a medical setting X-rays are created in an X-ray tube. These are located in diagnostic X-ray machines and on some linear accelerators. Components of an X-ray tube include the following:

- Tube housing: provides structure and protection for the rest of the elements.
- Cathode assembly or cathode block: the tungsten filament is the main part of the cathode.
- Anode: The anode structure is formed of the target usually made of tungsten which sits on an anode disk plus the stem rotor and stator. The anode can be stationary or rotating. Rotating anodes are most commonly used in diagnostic and therapeutic radiography.
- Tube envelope: the filament/cathode and target/anode are contained in the envelope, which provides vacuum, support and electrical insulation.
- Window/Port: usually made of beryllium. This allows the desired X-rays to leave the X-ray tube.

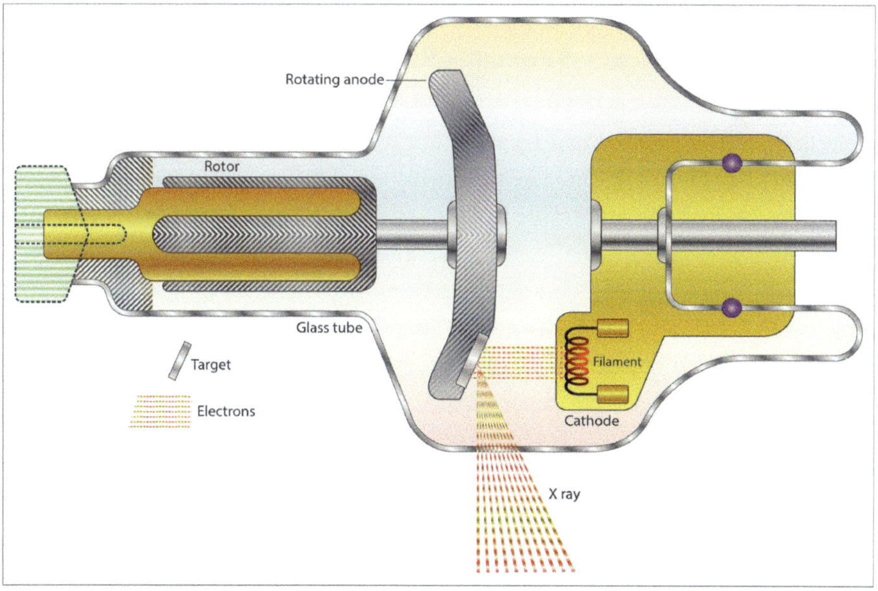

FIGURE 4.2 The (X-rays Produced By Interaction Accelerated Electrons Stock Vector [Royalty Free] 2232555593 | Shutterstock.)

The following steps outline how the X-rays are created:

1. **Electron Generation**
   - The filament in the cathode is heated by a low-voltage current.
   - Heat energy frees electrons from the tungsten atoms (thermionic emission).
2. **Electron Acceleration**
   - A high potential difference (measured in kilovoltage peak, kVp) is applied between cathode and anode.
   - Electrons are focused and accelerated across the vacuum towards the anode at very high speeds.
3. **Target Interaction**
   - When electrons strike the tungsten target, their kinetic energy is converted into:
   - Heat (~99%) – dissipated through conduction and cooling systems.
   - X-rays (~1%) – via two main processes (outlined below)

Bremsstrahlung Radiation (Braking Radiation)

X-rays are produced whenever high-energy electrons are suddenly slowed in a target. The energy lost by the particle may be emitted as a photon of bremsstrahlung radiation. The incoming (incident) electron passes close to the nucleus, and the attractive force of the nucleus decelerates the electron. The electron loses energy (i.e. it brakes), the closer to the nucleus the electron passes, the greater the deceleration and the greater the energy loss. If the electron is brought to a rest, maximum possible photon energy is produced. When the electron 'brakes', the bremsstrahlung X-ray photon is produced. A range of photon energies will be produced because different electrons will be decelerated to a greater or lesser extent due to their differing proximity to the nucleus.

Characteristic Radiation

The incoming (incident) electron ejects an inner shell electron from its shell. An electron in a higher shell drops down to fill the gap. As the electron drops down to fill the gap, a photon of characteristic radiation is then emitted. Unlike bremsstrahlung which produces a continuous spectrum, characteristic radiation produces a line spectrum of X-rays.

Factors effecting X-ray output:

1. Tube voltage
2. Tube current
3. Exposure time
4. Target material
5. Filtration

- **Tube Voltage (kVp)**
  - Controls the speed and energy of electrons.
  - Higher kVp = more penetrating beam, higher maximum photon energy.
- **Tube Current (mA)**
  - Controls the number of electrons emitted.
  - Higher mA = more photons produced, but energy remains the same.
- **Exposure Time (s)**
  - Longer exposure = more total photons.
- **Target Material**
  - Higher atomic number materials (e.g. tungsten, molybdenum) produce more efficient X-ray generation.
- **Filtration**
  - Removes low-energy photons that contribute to patient dose without improving image quality.

X-RAY INTERACTION WITH MATTER

When an X-ray beam enters the body, not all photons pass straight through to the detector panel (imaging) or exit the body. Instead, photons may be absorbed, scattered or transmitted depending on their energy and the type of tissue they encounter. These interactions determine both image contrast and patient dose.

1. **Coherent scattering (elastic or classical scattering)**
   This interaction occurs when an X-ray photon interacts with an atom but does not lose any energy. The incoming photon causes the atom's electrons to oscillate together. This occurs when the photon energy is small in relation to the ionisation energy of the atom. The photon is deflected as it passes the electrons and does not lose any energy, only changes direction.

2. **Photoelectric effect**
   The incoming X-ray photon must have an energy greater (or equal to) than the binding energy of the shell electron it is going to eject. The incoming photon is absorbed by an inner shell electron. All the photon energy above the binding energy is donated to the electron as kinetic energy (X-ray photon no longer exists). The bound electron is ejected from the atom as a photoelectron. The photoelectron will travel through material suffering successive encounters with other atoms, and eventually, when all its kinetic energy is exhausted, the photoelectron comes to rest and is captured by a nearby atom.

   The atom is left with an inner-shell vacancy. An electron from a higher energy shell transitions to fill this vacancy, resulting in the emission of characteristic radiation.

3. **Compton scattering**
   Occurs when the inbound photon energy is significantly higher than electron's binding energy. The photon gives up a proportion of its energy to the electron, the electron is ejected and the photon continues in different direction with lowered energy.

4. **Pair production**
   Occurs when photon energy is >1.02MeV, this is the energy equivalent of the mass of two electrons. The photon interacts with the electrical field of a nucleus and the photon is turned into a positron and a negatron (positive and negatively charged electrons). The negatron is slowed down; the positron will eventually collide with an electron resulting in annihilation radiation, i.e. mass is converted back to energy (in the form of two photons).

Clinical relevance:

- **Imaging:**
  - Photoelectric absorption underpins diagnostic contrast (e.g. bone appears white on X-ray images).
  - Compton scatter reduces image quality but can be managed with collimation, anti-scatter grids and digital image processing.
- **Radiotherapy:**
  - At high energies, compton and pair production dominate, allowing deeper penetration and dose delivery to tumours while sparing skin.

LINEAR ACCELERATOR PHYSICS

The linear accelerator or the linac as it's most commonly known, is the main piece of equipment that is used within radiotherapy. Linacs first started to emerge in the medical world in the 1950s and have advanced ever since with more compact linacs being introduced from the 2000s. Within radiotherapy, there is a variety of equipment utilised; however, the main focus of this section will be on the traditional linac. By traditional linac, this refers to the standalone arm-mounted linac as seen in Figure 4.3.

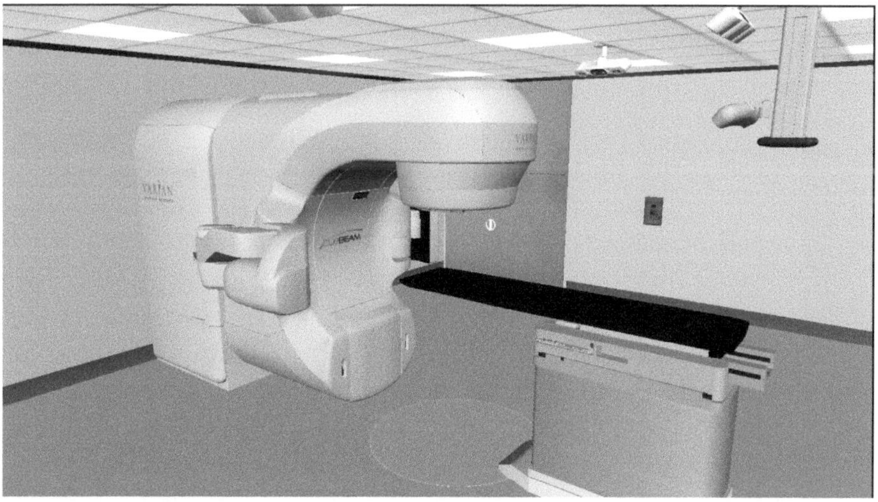

FIGURE 4.3 A traditional style linear accelerator (linac). (With permission from Vertual.)

This section will cover the design and operation of a linac, the function of the main component parts and discuss how the electron and photon beams are produced for the purposes of radiotherapy treatment.

The major components of a linac are:

- Electron gun – electrons are produced by thermionic emission from a heated cathode; an electrostatic field produced from the cathode cup focuses the electrons to a small area of the anode.
- RF generator – produces radiofrequency electromagnetic waves to enable the electrons to be accelerated in the accelerating waveguide. Either a klystron or magnetron can be utilised (these are discussed later).
- Ion pump – maintains the pressure in the vacuum system.
- Accelerating waveguide – accelerate the electrons, increasing their energy that can then be converted into X-rays or utilised as a clinical electron beam. Either a travelling or standing waveguide can be used depending on the manufacturers. It is essential that the electrons are correctly placed on the wave for maximum acceleration, which occurs in the first part of the waveguide, known as the bunching section. é1 receives the greatest acceleration therefore catches é2, é3 gains less acceleration therefore falls behind on the wave in relation to é1 and é2, resulting in the bunches merging together at the same point on the wave.
    - A standing waveguide accelerates electrons by a microwave generator (klystron or magnetron), sending radiofrequency (RF) energy into the first cavity. When the original travelling wave and the reflected travelling wave interfere with each other, a new wave pattern is formed: standing wave. There are fixed points called nodes that never undergo displacement. Midway between every consecutive node, there are points that undergo maximum displacement called antinodes. The wave produces regions of positive and negative charge that attract/repel electrons. Electrons entering a cavity at the optimum point in the wave cycle will find themselves accelerated. As the standing wave is a composite of two travelling waves, the resultant force on the electrons will be greater. As the nodes are fixed, alternate cavities always have a zero field, therefore never contribute to the acceleration of the

electron bunch. In practice, these can be moved to the side of the accelerating structure, making the overall length of the waveguide shorter.

- A travelling waveguide accelerates electrons by RF microwaves that propagate along the structure as a travelling wave. The RF power at the end of the waveguide can either be fed back to the input end or absorbed by an RF load. As the travelling wave moves down the waveguide, its electrical field induces a corresponding region of charge in the tube and copper discs.

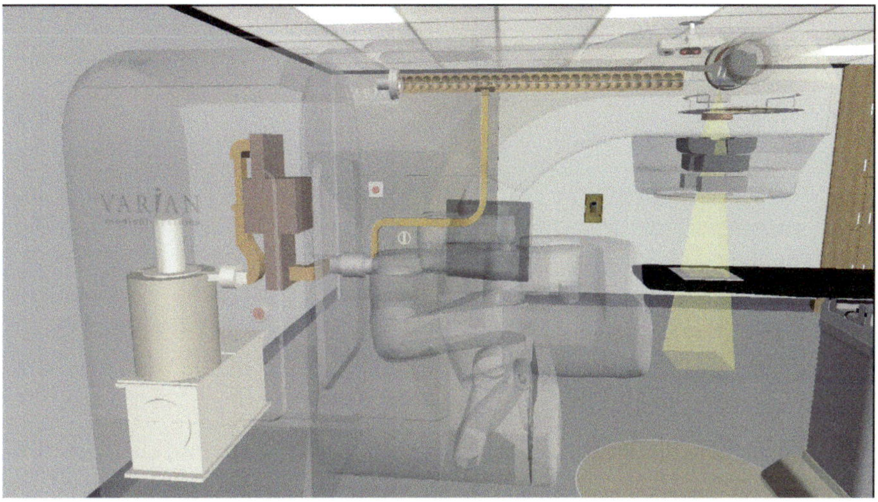

FIGURE 4.4 A waveguide within a linac treatment head. (With permission from Vertual.)

- Steering coils – minor imperfections in the accelerating waveguide can affect the position of the electrons; they are also influenced by external factors. Steering coils exert an electromagnetic force on the electron, keeping them in the centre of the accelerating waveguide. Paired electromagnets two on the x and two on the z, to control deviation. Set of coils at the beginning and end of the waveguide.
- Focussing coils – electrons have a tendency to diverge due to the RF wave and self-repulsion of a bunch of electrons. Focussing coils produce a magnetic field that aligns with the length of the waveguide; stopping the electrons from diverging.

- RF load – in a travelling accelerating waveguide, a RF load will absorb the RF wave.
- Treatment head – contains the bending magnets, target, ionisation chamber, optical system, collimators, wedges, scattering foils, beam flattening filter, multileaf collimators (MLCs).

Treatment head components:

- Target – Creates bremsstrahlung X-rays with a thin tungsten and rhenium disc. Remaining primary electrons are absorbed in a graphite absorber inside the target.
- Primary collimators – A block of high attenuating material that surrounds the target in the treatment gantry. It absorbs scatter from the target that is not directed into the treatment head.
- Beam flattening filter – Traditionally used to flatten beams or create uniform fields in conformal radiotherapy by reducing the dose rate due to attenuation. Made of lead, copper or steel.
- Scattering foils – Utilised in electron mode, to spread the beam to produce a uniform electron fluence across the treatment field. Made of thin high-Z metallic foil (e.g. lead or tantalum).
- Ionisation chamber – Dual system that monitors and controls dose rate, integrated dose and field symmetry ensuring consistent and safe output. It is a sealed unit, so it is not influenced by changes in temperature or pressure.
- Mirror – Thin Mylar mirror placed on the beam central axis under the dose monitor to enable patient set-up and show the position of the radiation beam.
- Secondary collimators – Consist of two independent jaws (X and Y), which allow the field generated by the linear accelerator to be shaped into a variety of rectangular shapes, with a typical maximum field size of 40 × 40cm.
    - Wedges – The linac uses the secondary collimators to create a wedge known as a dynamic wedge. They produce a profile similar to that of a physical wedge by varying the jaw position and/or output rate during treatment. When compared with physical

wedges, these non-physical wedges have several clinically relevant advantages, including reduction of treatment time, less scatter dose to peripheral areas and, in many cases, extended field size capabilities. Non-physical wedges inherently have no beam attenuation or beam hardening effect and thus offer more flexibility than physical wedges. The dynamic wedge generates a sloping dose distribution by moving one of the jaws with variable speed, while the opposite jaw remains steady. The rate the jaw moves is determined by the wedge angle required and is often indexed to the dose rate.

- Electron applicator – An attachment fitted to the treatment head and is available in several preset sizes. This applicator has several collimators which attenuate any lateral scatter. It usually extends to just several centimetres from the surface. The most distal aperture may have an electron cutout fitted. This allows the field to be shaped appropriately and is located in this position to prevent lateral scatter causing blurring of the beam edge. See Figure 4.7.

Magnetron

The magnetron is a device that produces microwaves. It functions as a high-power oscillator, generating microwave pulses of several microseconds' duration and with a repetition rate of several hundred pulses per second. The magnetron has a cylindrical construction, having a central cathode and an outer anode with resonant cavities machined out of a solid piece of copper.

The space between the cathode and the anode is evacuated. The cathode is heated by an inner filament, and the electrons are generated by thermionic emission. A static magnetic field is applied perpendicular to the plane of the cross-section of the cavities and a pulsed direct current (DC) electric field is applied between the cathode and the anode. The electrons emitted from the cathode are accelerated towards the anode by the action of the pulsed DC electric field. Under the simultaneous influence of the magnetic field, the electrons move in complex spirals towards the resonant cavities, radiating energy in the form of microwaves.

The generated microwave pulses are led to the accelerator structure via the waveguide. Typically, magnetrons power low-energy linacs (6MV or less).

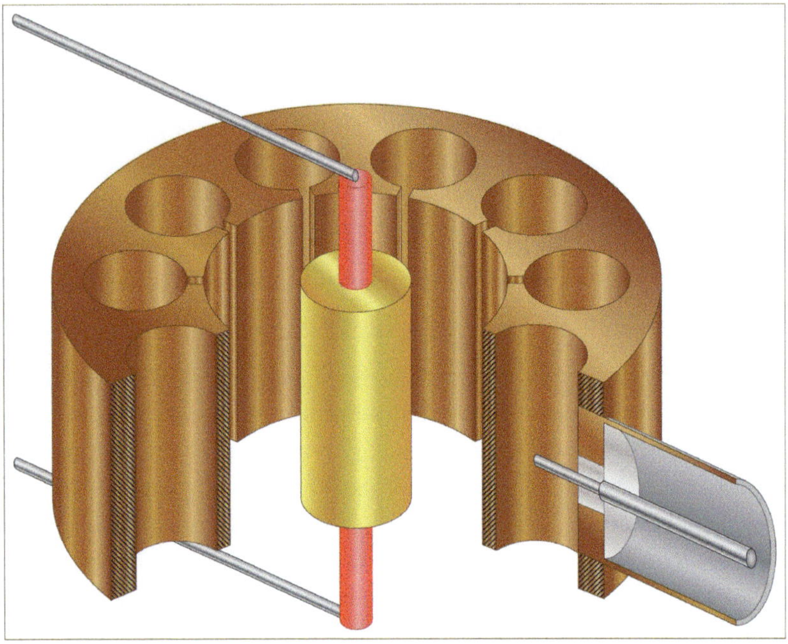

FIGURE 4.5 A magnetron. (Magnetron High-vacuum Tube Containing Cathode Anode Stock Vector [Royalty Free] 2326957527 | Shutterstock.)

Klystron

Operates in higher energy units and instead of being a source of microwaves, it is an amplifier of microwaves. It is driven by a source of low-energy oscillators. The electrons produced by the cathode are accelerated by a negative pulse of voltage into the first cavity, called the buncher cavity, which is energised by low-power microwaves. The microwaves set up an alternating electric field across the cavity. The velocity of the electrons is altered by the action of this electric field to a varying degree by a process known as velocity modulation. Some electrons are sped up, whilst others are slowed down, and some are unaffected. This results in bunching of electrons as the velocity-modulated beam passes through a field-free space in the drift tube.

As the electron bunches arrive at the catcher cavity, they induce charges on the ends of the cavity and thereby generate a retarding electric field. The electrons suffer deceleration, and by the principle of conservation of energy, the kinetic energy of electrons is converted into high-power microwaves.

Isocentre Concept

The isocentre is a fixed point within the treatment room around which the linac gantry, collimator, and treatment couch rotate. It is located at a standard distance of 100cm from the radiation source (source-to-axis distance, SAD).

Adjustable components that rotate around the isocentre:

1. The gantry
2. The collimator or treatment head
3. The treatment couch

The isocentre cannot be seen, but the lasers within the room replicate this position so that it can be visible when the lights are dimmed. The isocentre is where all three lasers (both lateral and sagittal lasers) coincide. The lasers are in a fixed position and are calibrated by the radiotherapy engineers to ensure they replicate the isocentre position, which can be measured at a distance of 100cm from the radiation source. Radiotherapy treatment is delivered at the isocentre, the isocentre is usually placed within the centre of the volume to be treated. To enable this to happen, the treatment couch can be adjusted. This is usually done by a reference point being selected, often this is where the permanent marks are placed on the patient. Isocentre shifts are then given, which move the patient from the reference marks to the isocentre. This is discussed in more detail in Chapter 3.

DOSIMETRY/RADIOTHERAPY TREATMENT PLANNING

Once the pre-treatment process is completed as discussed in Chapter 3, the CT planning dataset is sent electronically to the radiotherapy planning department also known as dosimetry. The radiotherapy treatment plan is created on the CT dataset, which will ultimately be sent through electronically to the treatment machines for the patient to start treatment. There are several steps involved in this process which will be discussed in more detail. To support this process there are reports created by the International Commission on Radiation Units and Measurements (ICRU), with the aim to create a set of standards that apply for different radiotherapy planning techniques that can be used internationally. This allows for standardisation of radiotherapy planning protocols, which are developed through research and best practice.

There are several factors which help a consultant decide which radiotherapy planning technique is the most suitable for each patient. The common ones are listed below.

1. The intended treatment site
2. Intended treatment outcome radical/palliative
3. If the patient has had previous radiotherapy
4. The patient's performance status
5. Patient choice
6. The organs at risk (OAR)
7. Tumour size, depth and position
8. The availability of equipment

Once the consultant has decided on these factors through the use of diagnostic imaging and patient consultations, a radiotherapy planning technique can be decided. The main ones are discussed in more detail below.

Electrons

Electrons are mainly used in the treatment of skin cancer due to the minimal long-term cosmetic effects when compared with surgery. The treatment area plus a margin is mapped out by an oncologist and therapeutic radiographer during a planning appointment. The shape, size and depth of the lesion are measured carefully. These details are then passed onto a mould room technician if the lesion is of an irregular shape, who then makes a customised electron endplate. These plates are attached to an electron applicator, which is affixed to the head of the machine. This apparatus not only helps to direct the beam, but the endplate also serves as a 'shield' to help protect the surrounding healthy tissue. In addition, bolus or build up may be required to create a homogeneous dose.

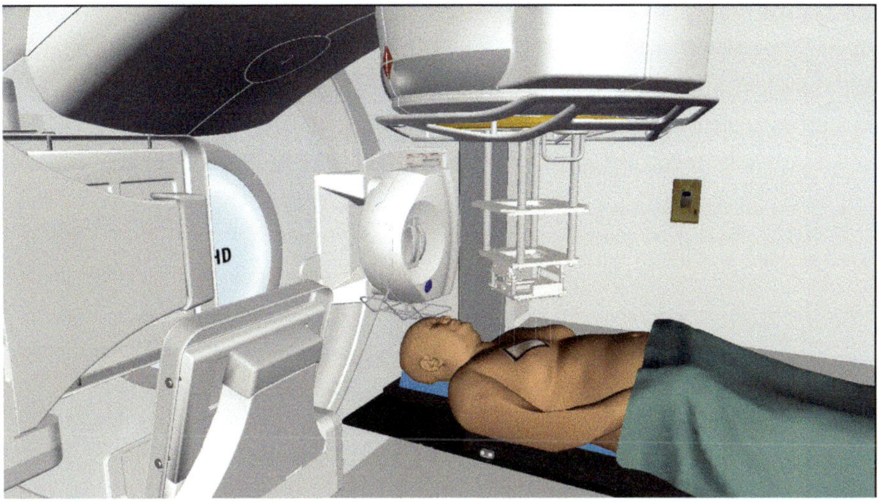

FIGURE 4.6 An electron beam applicator and end plate. (With permission from Vertual.)

Photon

Applied/direct/single field – This technique is a single field of radiotherapy, which can be treated at Dmax or a specified depth. Dmax refers to the depth of the maximum prescribed dose. The depth of Dmax is based on a 10 × 10cm field, depends on the energy of X-rays being delivered. The most common energies used are 6MV, where Dmax is around 1.5cm within the body and 10MV where Dmax is around 2.2 cm within the body. As mentioned above Dmax can also be prescribed to happen at a specific depth within the tissue; this can be factored into the monitor unit (MU) calculation. The patient will be given a higher amount of MU's to allow for the radiation to be deposited at a deeper depth due to attenuation and beam absorption in tissue. This technique can be used at any beam angle and is just one single field. This is usually used in the treatment of palliative patients to treat disease that is close to the skin surface such as spinal metastases, e.g. treating thoracic vertebra 5–6 with 8Gy in 1# at 4cm depth. The reason this is mostly used to treat palliative patients is due to the simplicity of the technique. There is minimal conformity and field sizes can be large. This technique requires no or minimal dosimetry; therefore, the training of staff is easier, and the process is quite short. Beam modifications can be used such as MLCs to shield any OARs, or hot spots. Overall, this technique is most commonly used to treat palliative patients, with the main aim being to reduce pain or symptoms, and is often referred to as conventional 2D radiotherapy. The 2D refers to the use of 2D imaging, which was used to support the localisation and planning process. Historically, the 2D imaging was produced by a simulator at pre-treatment, which would look similar to a linac but routinely took 2D X-ray images. This has now been phased out and replaced by the use of a CT scanner; however, due to the simple technique, it is still referred to as conventional radiotherapy.

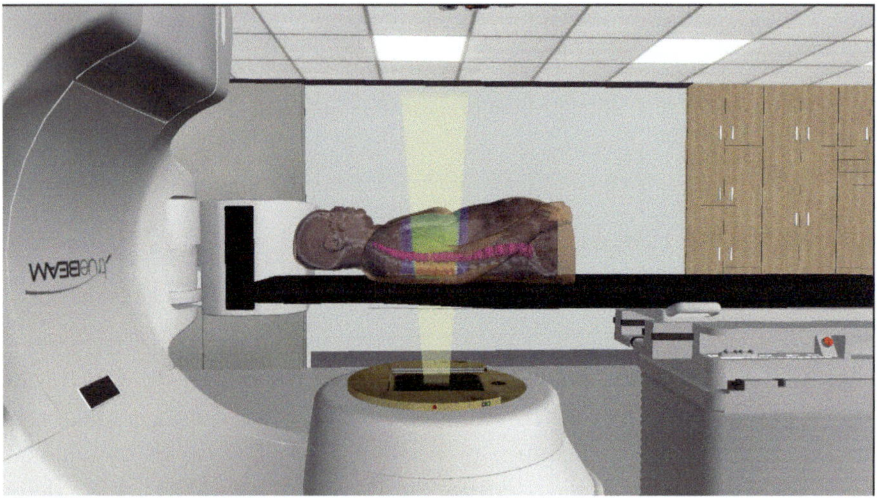

FIGURE 4.7 An example of a single posterior field with dose distribution. (With permission from Vertual.)

Parallel opposed pair (POP) – This technique consists of two opposing beams (180 degrees apart), where the dose is prescribed to mid-plane/mid-separation. This technique can be used at any beam angle, but common ones are 0 and 180, or 90 and 270 degrees. This is usually used in the treatment of disease that is seated deeper within the patient or would benefit from two beams to increase the dose such as treating the whole brain or hip. Again, just like applied fields, this technique is usually used in the treatment of palliative patients, due to its simplicity, minimal conformity and field sizes can be large. This technique requires no or minimal dosimetry; therefore, the training of staff is simple, and the process is short. Beam modifications can be used such as MLCs to shield any OARs, or hot spots. Overall, this technique is most used to treat palliative patients, where the main aim is to reduce pain or other symptoms.

Three-dimensional Conformal Radiotherapy (3DCRT)

Delivers multiple radiation beams at various treatment angles, routinely around 3–4 beams. These beams all coincide at the isocentre, therefore is also known as an isocentric technique. The three-dimensional (3D) refers to the use of 3D CT imaging, which provides a volumetric dataset of the patient's anatomy. This dataset is used in the treatment planning system to contour structures and calculate dose distributions, creating a visual map of the dose. This allows the planners to visually see and adjust the dose to ensure it is meeting departmental protocols. As the name refers, conformal radiotherapy means that the plan is conformed to the treatment area and uses beam modifications or beam shaping such as wedges and MLCs to conform the dose to the tumour. This enables healthy tissues to be spared but as much dose to the treatment area itself. When radiation is delivered, it assumes the area is smooth and flat therefore beam modifications such as wedges account for the human body where the contour isn't smooth and flat. It can also help to adjust the dose distribution where tumours are of varying shapes and sizes too.

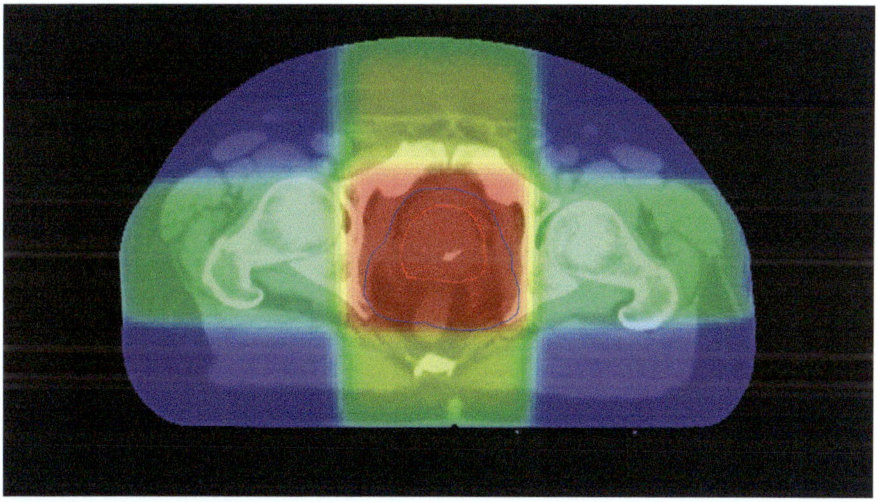

FIGURE 4.8 The dose distribution on a 3DCRT technique. (With permission from Vertual.)

Intensity Modulated Radiotherapy (IMRT)

This technique is relatively new and became more standardly used in last 10 years or so. This technique consists of multiple beams, delivered at various treatment angles, routinely between 7 and 9 beams. The difference with this technique is that it allows for modulated beams. This means the intensity of the beam could be changed during the delivery of the beam. Therefore, further enabling a higher dose to be delivered to the tumour and less dose to surrounding healthy tissues. This developed the radiotherapy world as it meant dose escalation could be achieved, as even higher prescription doses could be used as they were able to be confined to the treatment area. Beam modifications such as MLCs were used, allowing for shaping of the beam but also modulation of the beam. MLCs could be brought into the treatment field which would shape and modulate fluence, therefore, helping to reduce the dose in the treatment area for certain parts of the field. This technique could be delivered in two different ways: step and shoot or dynamic. Step and shoot refers to the beam being delivered and then the beam would pause while the MLCs readjusted their position and then the beam would carry on. However, dynamic sliding window means the beam is continuously delivered at that fixed angle but the MLCs would be moving all the time. Step and shoot is being phased out, and dynamic IMRT is the most common. This technique is faster as the beam does not need to interrupt to allow for the repositioning. This technique of radiotherapy planning is very intricate and time consuming, meaning staff require advanced training.

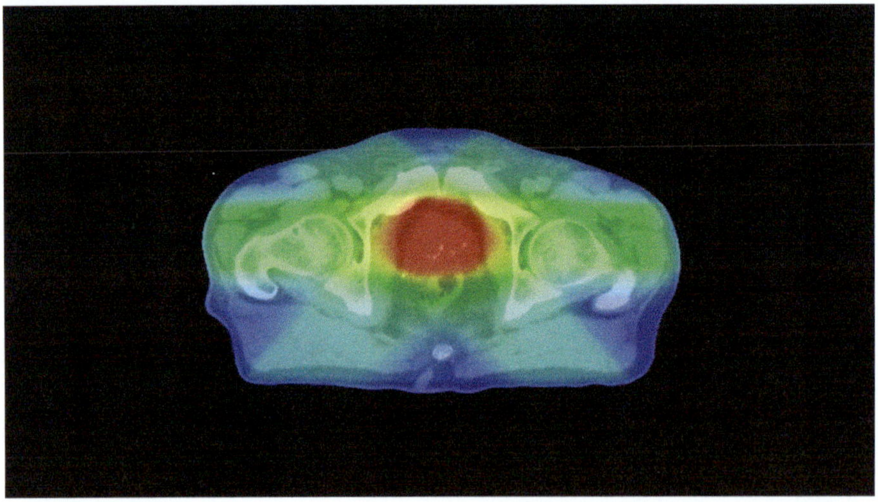

FIGURE 4.9 The dose distribution on an IMRT technique. (With permission from Vertual.)

Volumetric Modulated Arc Therapy (VMAT)

This technique is a further advancement on the IMRT technique. The only change with this technique is that the linac can rotate while delivering the treatment. Therefore, there are no longer fixed fields; they are now referred to as arcs. Treatments can differ from half an arc to a full arc or multiple half/full arcs depending on the treatment plan. MLCs are still used to modulate and shape the beam, and there is no real need for the used of wedges as the same can be achieved with the MLCs.

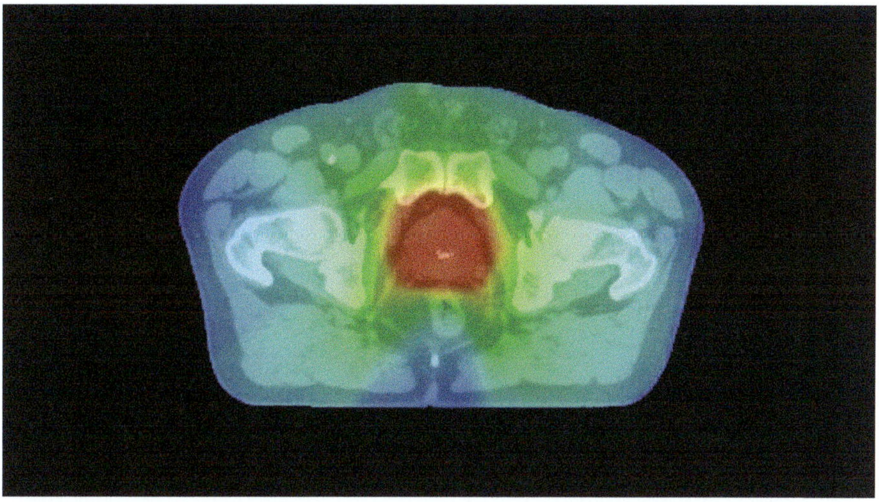

FIGURE 4.10 The dose distribution on a VMAT technique. (With permission from Vertual.)

Inverse Planning

Another change that was brought in to support IMRT/VMAT planning is inverse planning. Unlike the other planning techniques where the planner/dosimetrist would make all of the adjustments to create the plan they required; inverse planning relies on the planner telling the system the aims of what it would like to achieve from the plan and then the system would use optimisation algorithms to meet the dose objectives. This further supports the need for extensive staff training. This technique was firstly associated with complex treatments where very tight margins are required, due to the proximity of the tumour to healthy structures such as in head and neck tumours. Over the years, this technique has become more routinely used for the majority of radical treatments, and for the majority of treatment sites.

Radiotherapy Planning Volumes

As previously mentioned, ICRU helped to create internationally acceptable recommendations for radiotherapy. One of the reports used to support radiotherapy planning is the ICRU 50 report, which was published in 1993 (ICRU 1993). This report focuses on the prescribing and reporting of radiotherapy by describing parameters of volumes and doses. This was then supplemented by the ICRU 62 report in 1999, which further develops this, but also focuses on the concept of OARs (ICRU 1999). These reports focus on 3DCRT planning, therefore in 2010, the ICRU 83 report was published with a specific focus on IMRT (ICRU 2010). However, the ICRU 50 and 62 supplement reports identified volumes that should be created when radiotherapy planning which are still relevant when planning any technique, these are:

- Gross tumour volume (GTV) – The gross palpable or visible/demonstrable extent and location of malignant growth.
- Clinical target volume (CTV) – A tissue volume that contains a demonstrable GTV and/or subclinical microscopic malignant disease, which has to be eliminated.
- Planning target volume (PTV) – Is a geometrical concept and is defined to select appropriate beam sizes and beam arrangements, taking into consideration the net effect of all possible geometrical variations, in order to ensure the prescribed dose is actually absorbed in the CTV. Dose variation between +7% and −5% of the prescribed dose is generally accepted. Composed of two margins: internal and setup.
    - Internal margin (IM) – Accounts for anatomical and physiological variations such as small brain movements, large and predictable respiratory or cardiac pulse, large and unpredictable rectal and bladder filling. Therefore, these can lead to interfraction motion, which is motion between fractions and/or intrafraction motion which is motion during a single fraction.
    - Set-up margin (SM) – Accounts for variations in patient position and alignment with beams. These uncertainties can depend on variations in patient position, mechanical uncertainties of equipment, dosimetry uncertainties, set-up errors, data transfer errors, human related errors.
- Treated volume (TV) – Volume enclosed by an isodose surface, selected and specified by the radiation oncologist as being appropriate to achieve the purpose of treatment. Usually for tumour control, this would be the 95% isodose.

- Irradiated volume (IV) – Tissue volume which receives a dose that is considered significant in relation to normal tissue tolerance, this can be represented by the isodose that represents 20Gy.
- Organs at risk (OAR) – Organs at risk are normal tissues whose radiation sensitivity may significantly influence treatment planning and/or prescribed dose.
- Planning organ at risk volume (PRV) – Any movements of the OAR or uncertainties of set-up may be accounted for within this margin.

Dose Volume Histogram

A way for the dose to be shown within these volumes is using a dose volume histogram (DVH). Volumes will be contoured on the radiotherapy treatment planning system (TPS), then these can be evaluated using the DVH which will show how much dose each structure is receiving. The planner can then amend the plan to ensure the dose meets the constraints for each volume.

Below gives a simplified outline of the steps involved in creating a VMAT Prostate plan; each department will have their own instructions therefore may differ slightly.

1. Electronic CT dataset received from pre-treatment localisation appointment.
2. Dataset is reviewed to check appropriate anatomy is within the dataset and the dataset can be labelled as per departmental protocol such as patient ID number and scan date.
3. A patient contour is added (some updated TPS add this automatically). This is an essential step as the TPS now knows where the patient is within the dataset. If this is not done automatically, this can be completed by a planning radiographer/dosimetrist.
4. Next, the consultant will review the CT scan and contour appropriate volumes. Standardly for a prostate radiotherapy plan, the prostate will be contoured and considered the GTV. This volume can then be expanded to create the CTV which includes the prostate and at least the base of the seminal vesicles (depending on staging). The PTV can be automatically created from the CTV, where the CTV-PTV is grown 10m isotropically, the posterior margin can be reduced to 5mm if it includes too much rectum. Additional diagnostic imaging can be

used such as MRI to support this process and can be fused or registered with the planning CT scan. The consultant will review these margins and adjust any they feel need amending.

5. OAR will then be contoured; this can be completed by the consultant and/or the planning radiographers/dosimetrists. The consultant should always review the OAR contouring if not completed by themselves. OARs routinely contoured for a prostate plan are the rectum, bladder, penile bulb and femoral heads.

6. The planning radiographer/dosimetrist will now create an inverse planned prostate plan. Objectives will be set outlining the planning aims, which the TPS will then try to achieve. A prescription will also be included, the most common is 60Gy in 20 fractions over 4 weeks for prostate treatment. The main radiotherapy aim is to deliver the prescription dose to the PTV and minimal dose to the OAR. However, there will always need to be a balance as some OAR are so close or within the PTV that a compromise is needed. This is where the consultant will decide on what is achievable and what the priorities are.

TABLE 4.1 Objectives that Are Commonly Used for the PTV

Criteria	Dose Volume Constraint	60Gy/20f
Minimum D98%	≥95% prescription dose	≥57Gy
Median D50%	99% to 101% prescription dose	59.4-60.6Gy
Maximum D2%	<105% prescription dose	63Gy

Source: ICRU (2010).

TABLE 4.2 A Clinical Example of Commonly Used Constraints for Prostate OARs with a Prescription of 60Gy/20#/4 Weeks

OAR	Prescription Dose 3Gy/#	Volume
Bladder	V40Gy	<50%
	V48Gy	<25%
	V56.8Gy	<5%
	V60Gy	<3%
Femoral Heads	V40Gy	<5%
Bowel	V36Gy	<78cc (cubic centimetre)
	V40Gy	<17cc
	V44Gy	<14cc
	V48Gy	<0.5cc
	V52Gy	0cc
Penile bulb	V40Gy	<50%
	V48Gy	<10%

Top Tip: how to read Table 4.2. Example: V40Gy <50% means volume of the bladder receiving 40Gy needs to be less than 50%.

7. Once the planner/dosimetrist feels they have achieved the appropriate plan, the consultant will review the plan. This may not be done immediately depending on consultant availability.
8. If the consultant feels the plan is acceptable the plan will be signed off for treatment. If not, adjustments will be made until an acceptable plan is achieved.
9. Then plan is checked by physics to ensure all dose reporting is accurate.
10. The plan is electronically sent through to the preparation area where the treatment radiographers will complete checks prior to the patient starting treatment. The planning process is completed at this stage.

REFERENCES

International Commission on Radiation Units and Measurements (ICRU) 1993. ICRU Report 50: Prescribing, recording, and reporting photon beam therapy. https://www.icru.org/report/prescribing-recording-and-reporting-photon-beam-therapy-report-50/ (accessed 17 December 2025).

International Commission on Radiation Units and Measurements (ICRU) 1999. ICRU Report 62, prescribing, recording and reporting photon beam therapy (Supplement to ICRU 50) – ICRU (accessed 17 December 2025).

International Commission on Radiation Units and Measurements (ICRU) 2010. ICRU Report 83, prescribing, recording, and reporting intensity-modulated photon-beam therapy (IMRT) – ICRU (accessed 17 December 2025).

CHAPTER 5

Pre-Placement Information

INTRODUCTION TO PRE-PLACEMENT INFORMATION

Preparing for clinical placement is a really important step; this will help you feel confident and equipped for the challenges you may face. Before commencing your clinical placement, it is important to know what will be expected of you. This chapter will provide relevant information to support your transition into the clinical department. It will give examples of the different communication styles and when to apply them, depending on who you are communicating with. It will cover the day-to-day running of a radiotherapy department, identifying the different skill mixes and their roles and conclude by outlining the key aspects of radiation safety.

PROFESSIONALISM

Professionalism is a term you may hear a lot throughout your time studying therapeutic radiography. When the term professionalism is used, what is being referred to is the way in which you conduct yourself. Therefore, this links to the behaviour you demonstrate and your attitude towards your work, including how you engage with colleagues, patients and carers. Acting in a professional manner helps to earn the trust and respect of colleagues, patients and carers. There are six main areas which all help to demonstrate professionalism.

1. Competence – having the necessary skills, knowledge and ability to perform your role.

DOI: 10.1201/9781003639183-5

2. Accountability – taking responsibility for your actions.
3. Integrity – acting with honesty, fairness and with strong moral principles.
4. Respect – treating colleagues, patients and carers with consideration.
5. Reliability – being dependable and consistent in completing responsibilities and commitments.
6. Appearance and communication – presenting yourself appropriately and communicating clearly (this is expanded on in the communication section).

These six areas are developed throughout your time within the academic and clinical environment; however, there is an expectation that you have a basic understanding of professionalism and what is expected of you. This is especially valid for when you start your practice placement. Patients are in an extremely vulnerable position while undergoing radiotherapy; therefore, the need for a basic understanding of professionalism when in a clinical department is essential. With this in mind, your professional suitability is always monitored by your practice educators and academic team. Professional suitability is used to monitor how well your behaviour aligns to the university and clinical departments' standards. These are developed from the HCPC Standards of Conduct, Performance and Ethics and the Society of Radiographers (SoR) Code of Professional Conduct. It is appreciated that professionalism develops with experience; therefore, this is something that will continue to develop even once you become a fully qualified therapeutic radiographer. These requirements are needed to ensure the safety of colleagues and patients. Below are examples of how professional suitability can be demonstrated:

- Effective communication in professional relationships
- Professional demeanour in appearance and behaviour
- Mutual respect in professional relationships and in the needs and priorities of the working environment
- High standards of timekeeping and attendance
- Ability to reflect on practice to strive for the best

Professional unsuitability is demonstrated through any acts or omissions which could be judged to endanger public safety or bring the profession into dispute. Below are some examples of how this can be represented:

- Failure to comply with professional code of conduct
- Poor attendance and/or timekeeping
- Consistent failure to communicate appropriately in a manner which impacts service delivery
- Failure to demonstrate consistent application to the development of professional skills
- Failure to exercise due consideration for the safety and welfare of the patient
- Any action leading to a justified formal complaint
- Any action leading to a disciplinary procedure

The examples above for both suitable and unsuitable professionalism are not an exhaustive list and are based on individual behaviours but should give an idea of the desirable behaviours that need to be demonstrated to thrive and succeed in the clinical environment.

Top tip: Regularly ask for constructive feedback from mentors and practice educators on your professional suitability. As mentioned earlier, professionalism is something that you will continue to develop; therefore, the more feedback you can receive, the more you can apply to become the therapeutic radiographer you want to be.

Top tip: Completing reflections when tasks go well or not so well will help you develop your professionalism. Take notes of the good practices you witness from other therapeutic radiographers to work on for yourself.

UNIFORM

As listed above, as an area to demonstrate professionalism, point 6; appearance and communication – presenting yourself appropriately and communicating clearly. This shows how important it is that therapeutic radiographers present themselves in a clean and tidy manner, as this helps others regard them in a trustworthy and respectful way. This is supported by the need for a uniform, which ensures therapeutic radiographers are dressed appropriately to carry out their role, that they look professional and it ensures all therapeutic radiographers are identifiable. The chosen uniform may vary from department to department; however, all therapeutic radiographers within one department will wear the same uniform. Some departments opt for scrub-type uniforms and others more traditional tunic and trousers/tunic dress which will be provided. It is important that uniforms are maintained appropriately to ensure they continue to look professional and that they are cleaned after every use. Not only does this help to ensure therapeutic radiographers look clean, tidy and ready for duty but it also limits the spread of any infections. Other than the uniform that is provided, something to consider is appropriate footwear. It is important to note that therapeutic radiography is a physical job and comfort is necessary. The usual requirement is black leather shoes, which are closed in and can be wiped clean. Each department will have their own uniform policy. Some of the requirements that can be expected for health and safety reasons and to maintain hygiene standards are highlighted below.

1. Jewellery – simple stud earrings and one plain wedding band.
2. Bare below the elbows – watches, braceletsand cardigans not to be worn when in the clinical environment. **Top tip**: a fob watch can be attached to the uniform and are usually permitted.
3. Plain, dark coloured socks to be worn under trousers.
4. Hair below the shoulders to be tied up and off the collar.
5. No nail varnish or false nails to be worn.
6. Avoid the use of strong perfumes.

Another requirement is that uniforms should not be worn on public transport, again to limit the spread of infections.

The way in which we present ourselves is heavily linked to communication.

COMMUNICATION

Communication is a two-way process where a message is transmitted between two or more people with the intent of creating a shared understanding. It is a fundamental aspect of professional practice in therapeutic radiography. As a student, demonstrating the ability to 'communicate appropriately and effectively' is outlined in the HCPC guidance on Conduct, and Ethics for Students (HCPC 2024). Therefore, it is essential that you work on developing and refining your communication skills. Each clinical placement is a valuable opportunity to practice and strengthen your skills.

During your clinical placement, you will be expected to communicate clearly, appropriately and professionally, not only with your peers and academic staff, but also with patients, their families and carers and members of the wider multidisciplinary team. It is important to remember that each interaction requires you to demonstrate; professionality and understanding.

Why Is Communication Important?

At its core, communication is the process through which we exchange information. In healthcare, this exchange underpins safe practice, effective teamwork and collaborative learning. Effective communication also goes towards building trust between patients and caregivers. Miscommunication can lead to errors and compromise care outcomes. Communication comes in many forms and is both verbal and non-verbal. Effective communication can be achieved through the use of each type during every interaction.

Forms of Communication

1. Non-verbal:
   One key aspect of non-verbal communication is open body language. This refers to a relaxed, approachable posture that signals attentiveness, confidence and empathy. An example of open body language would be uncrossed arms and appropriate eye contact. Facial expressions are also a big aspect of successful communication. Maintaining appropriate eye contact and a neutral yet friendly expression conveys that you are listening to that person and are sincere in your response. Conveying a message with a frown, pursed lips or an eye roll will come across as uncaring, unsympathetic and unprofessional. Written communication is also extremely important. This form of communication is used throughout all aspects of the radiotherapy pathway, including audits, correspondence and incident reporting. The correct documentation of information is imperative for maintaining patient safety and is sometimes the only

option available to pass on important information clearly whilst maintaining confidentiality.

2. Verbal:
   The appropriate use of language is essential for maintaining professional boundaries in therapeutic radiography. Tailoring your language to suit an individual's needs can be very effective. For example, some patients who are not familiar with medical jargon and slang, tailoring your language not only demonstrates your ability to communicate complex information clearly but also creates a respectful and professional atmosphere. In addition, tone of voice plays a crucial role. This factor can be used to convey reassurance and understanding.

   Both verbal and non-verbal communication require continuous practice. To further efficiency, these skills must be coupled with active listening. This ensures that the correct information is being relayed in a manner that everyone can understand.

3. Active listening:
   Active listening is a core component of effective communication. It requires your full attention, appropriate responses (verbal and non-verbal) and clarification when necessary. Essentially, this is a process in which you 'repeat back' information that is being given to you, to ensure that not only you understand, but also to demonstrate that you are listening.

Active Listening Example

A patient receiving radiotherapy for prostate cancer expresses anxiety before their first treatment session, saying, 'I'm really nervous about what's going to happen today'.

Active Listening Response from the therapeutic radiographer:

> I can hear that you're feeling really anxious about your first treatment, and that's completely understandable. Would it be helpful for me to explain the steps of the treatment again and talk through any concerns you may have?

Why this is active listening:

- The radiographer acknowledges the patient's feelings without dismissing them.

- They reflect back the emotion ('you're feeling really anxious').
- They offer support by giving the patient an opportunity to ask questions or seek reassurance.

Barriers to Effective Communication

Despite best intentions, several hurdles can impede communication, especially in a clinical setting. Therapeutic radiography is an extremely diverse setting; therefore, you may come across a patient who is visually impaired and therefore not able to read consent, deaf and unable to hear you or may not speak or understand English. However, every barrier has a solution. It is essential that you are aware of the processes in place in your clinical placement to ensure inclusiveness. Another barrier to be aware of is bias, both from yourself and others. Bias is the action of opposing or supporting something unfairly and accumulates into stereotypes. They can be innate or learned and may result in unprofessional or dangerous practices. Talking loudly and slowly is a reflex many people have when talking to someone with dementia or learning difficulties as it is assumed they will not understand complex information. These actions can result in that person feeling shame, anger and patronised, therefore, compromising the communication pathway.

Before engaging in any form of communication, it is important to note that both the environment in which a conversation occurs and the emotional state of the individual play a significant role. For example, if a patient is feeling anxious and is spoken to in a busy, crowded area, they are less likely to fully process or retain the information being given. In contrast, delivering the same information in a calm, private setting can help the individual feel more at ease and therefore encourage engagement and understanding.

How to Introduce Yourself

As a student, you will spend a lot of time introducing yourself. This is a skill that requires both verbal and nonverbal communication and is an imperative part of being a professional. Firstly, present yourself in a professional manner. When introducing yourself, use open body language. If possible, try not to fiddle with your uniform or an object. An easy way to achieve open body language is to put your hands behind your back and look at the person you are going to talk to.

Secondly, the best way to introduce yourself is by saying, 'Hello my name is….and I am a student therapeutic radiographer'. The tone of your voice should be calm and even, at an appropriate decibel. It is helpful to have a 'script' in your head that is specific to a certain situation. For example, when talking to a patient for the first time;

- Hello my name is....
- I am a student radiographer
- I will be assisting the radiographers with your treatment/first day chat/making your mask...today

When talking to a staff member for the first time;

- Hello my name is....
- I am a student from [insert university name] in [insert year of study]
- My aims for this placement are.......
- This can be achieved by....

This is a simple interaction that can be very impactful as it will help to build trust and rapport.

The next section aims to help you understand the different roles of staff and who may be able to support your learning.

See Chapter 9 for more conversation starters and communication tips including first-day chats.

STAFF ROLES

Within the therapeutic radiography profession, there are many different roles and responsibilities. Typically, a newly qualified therapeutic radiographer will take on the role of a band 5 therapeutic radiographer. This usually starts as a rotational treatment therapeutic radiographer. This means that they rotate around different areas within the department, following a rota created by the department. However, it is important to note that the therapeutic radiography profession is changing, and now more newly qualified radiographers are going straight into roles in specific areas such as dosimetry planning and pre-treatment. Not only is the profession changing, but it is also expanding, there are a lot more specialist roles and more career progression than there ever has been, some have been identified below.

Band 6 specialist roles:

- Practice educator – key focus on supporting students in their day-to-day clinical placement.

Band 7 specialist roles:

- Site-specific review radiographers examples, include paediatric, CNS (central nervous system), thorax – advanced specialist knowledge of the particular treatment site they are looking after who support patients through the treatment.
- Information and support radiographer – offer patients advice and support throughout their treatment with a range of issues such as helping with financial issues to side effects support.
- Brachytherapy lead radiographer – advanced knowledge of brachytherapy and the treatment pathway.
- Research radiographer – advances therapeutic radiography practice through scientific research such as clinical trials, quality improvement programmes and quality assurance.

Band 8 or above specialist roles:

- Treatment superintendent radiographers – advanced knowledge and experience of radiotherapy treatment, able to support treatment radiographers through decisions on the treatment units.
- Quality and safety superintendent radiographers – responsible for ensuring the department is maintaining quality standards and has the right policies in place to ensure safety.
- Education lead/practice development lead – key focus on students and staff development.
- Consultant radiographer – advanced knowledge on particular tumour sites, able to review patients and support the planning of the treatment.
- Radiotherapy service manager – responsible for the running of the radiotherapy department.

This is not an exhaustive list, and it can vary between departments.

There are four main areas within the radiotherapy department which are listed below.

1. Pre-treatment
2. Dosimetry planning
3. Preparation
4. Treatment

Some departments may have additional areas such as brachytherapy, superficial and/orthovoltage units. Within these areas, there can be a variety of therapeutic radiographers with different roles and responsibilities; however, this will vary between departments.

To ensure departments have the appropriate skill mix, there are a variety of roles which relate to experience and seniority. Below these are outlined:

- Band 5 Therapeutic Radiographer – When a therapeutic radiographer qualifies, this is the role they are now qualified to undertake.
- Band 6 Therapeutic Radiographer – This usually relates to the therapeutic radiographer becoming a senior member of staff. The timeframe for this is usually around 1.5 years of experience; however, this does vary.
- Band 7 Therapeutic Radiographer – The next role is a band 7 team leader role. This indicates that the member of staff is a senior member of the team and is a team leader, therefore has overall responsibility for that unit during their shift. The timeframe for this is usually around 3 years qualified; however, this can vary quite vastly too.
- Band 8 Therapeutic Radiographer – usually known as a superintendent radiographer.

Departmental Skill Mix and Shifts

Depending on the busyness of departments, it can depend on which shifts are organised. Radiotherapy departments are usually staffed from 8 am to 6 pm Monday to Friday, but shift times can vary until 8 pm. This normally means that departments operate with two shift patterns, known as an early shift and a late shift. Each shift will have two therapeutic radiographers, one band 5 and one senior member of staff, either a band 6 or a band 7. The working day is 7.5 hours, plus an unpaid lunch break. Often, a 20-minute tea break is given at good will. Some departments can operate with an extra team member who can oversee completing any administration tasks, helping with patient queries and the smooth running of the unit.

Looking at the individual areas below, an outline of the skill mix is identified.

1. Pre-treatment
   This area usually consists of band 5, band 6 and band 7 therapeutic radiographers. These therapeutic radiographers can be permanent pre-treatment radiographers, meaning they are permanently based within the pre-treatment area. Or they can be rotational therapeutic radiographers, meaning they rotate through the treatment floor and pre-treatment, or it can be a mix of both. The main thing is that there needs to be a skill mix of junior and senior therapeutic radiographers to ensure the team has enough experience to deal with all situations. Within pre-treatment, there is usually a band 8 therapeutic radiographer often known as a superintendent. This again can be permanent pre-treatment or a rotational therapeutic radiographer. Some pre-treatment departments have sub-sections within the area to cover other tasks such as virtual simulation, where therapeutic radiographers can help create simple radiotherapy plans in conjunction with a doctor. Again, it may be permanent or rotational therapeutic radiographers that rotate through these areas.

2. Dosimetry planning
   This area of the radiotherapy department runs similarly to that of pre-treatment, where there can be a mix of permanent therapeutic radiographers or rotational radiographers who are able to create radiotherapy plans. Some hospitals also run the dosimetry planning department completely separately to the radiotherapy departments; this depends on the hospital. There are other roles within this department such as physicists who are not therapeutic radiographers but work closely with them. If therapeutic radiographers are working within the department, this operates with a similar skill mix of band 5, 6 and 7 therapeutic radiographers. Plus, a band 8 superintendent who may be a therapeutic radiographer or

physicist. There are often other leads within this area such as a lead physicist.

3. Preparation
   This area is where all the checks are completed on radiotherapy treatment records and plans before they are sent through for treatment. Treatment prescriptions and doses are double-checked to ensure there are no errors prior to sending. This can be done online or on paper depending on the department. Most departments are working towards being paperless if they are not already. The skill mix within this area can vary, but there is usually a mix of band 5, 6 and 7 therapeutic radiographers. Again, there is usually a superintendent that oversees the area too.

4. Treatment
   This area consists of rotational therapeutic radiographers, who rotate around all of the different treatment units. As discussed earlier, there are usually two shift patterns, an early and a late shift. The skill mix consists of band 5, band 6 and band 7 therapeutic radiographers, with a band 5 therapeutic radiographer being paired with a senior band 6/7 member of staff. There will then be at least one band 8 superintendent treatment radiographer to oversee all of the treatment units.

RADIATION SAFETY ON PLACEMENT

Ionising radiation can have detrimental effects on human health such as cancer induction. Therefore, it is imperative that you follow radiation safety protocols and procedures whilst on placement. Radiation protection ensures the safety of patients, radiographers, other staff and the public, protecting them from the biological effects of ionising radiation.

There are two important legislations in place that regulate the use of ionising radiation. The Ionising Radiations Regulations (IRR) 2017 and Ionising Radiation (Medical Exposure) Regulations IR(ME)R 2017. IRR is concerned with occupational exposures and the radiation protection of the public (Legislation.gov.uk 2017a, 2017b). IR(ME)R protects any individual undergoing medical exposure, in this case patients undergoing radiotherapy (Legislation.gov.uk 2017a, 2017b). You will hear about IR(ME)R frequently in your studies, and you will work under this legislation as 'an operator' when you qualify (IR(ME)R 2017). Below is a summary of what is included in the regulations.

IR(ME)R

Key areas covered in IR(ME)R include:

- Ensuring that people working with radiation know their role and responsibilities
- Easy identification of responsibility
- Ensuring that practitioners and operators have adequate training
- Justification and optimisation of doses and the principles of keeping doses ALARP

Under IR(ME)R there are five roles:

1. Employer
2. Referrer
3. Practitioner
4. Operator
5. Medical physics expert

The employer can be the NHS trust; they are responsible for creating a framework for staff to work within, which is done through written procedures and protocols. One of these protocols is the local rules.

The referrer could be a GP or a consultant; they need to supply the practitioner with sufficient medical data (such as previous diagnostic information or medical records) relevant to the medical exposure requested by the referrer to enable the practitioner to decide on whether there is a sufficient net benefit.

The practitioner is responsible for the justification of a medical exposure. A clinical oncologist can take the role of practitioner in a radiotherapy context.

Therapeutic radiographers work as operators carrying out all practical aspects of medical exposure. As well as delivering the radiation, the practical aspect includes identifying the patient for treatment, checking pregnancy status and image manipulation.

Medical physics experts help to develop and implement protocols for the use of radiotherapy. They also oversee the quality assurance of equipment and are involved in the installation of new equipment. Medical physics experts usually have a physics degree or work in clinical physics.

Local Rules

On arrival at the radiotherapy department for placement, you will be asked to read the local rules before you start your placement. The local rules should outline (but not limited to) the following:

1. Name(s) of the radiation protection supervisor (RPS) and radiation protection advisor (RPA) (required by IRR17). The role of the RPS includes monitoring radiation levels, ensuring proper use of protective equipment and maintaining safety records. The role of the RPA includes supervising implementation of radiation safety legislation, investigates radiation incidents and oversees new equipment installation.

2. Identification, description and boundaries of controlled and supervised areas (required by IRR17). A controlled area is an area wherever a person is required to follow special procedures to restrict significant exposure. For example, this could be the treatment room itself. It is normally marked with a radiation warning symbol.

FIGURE 5.1 Radiation symbol. (Iso 361 International Ionizing Radiation Trefoil Stock Vector [Royalty Free] 2425012801 | Shutterstock.)

A supervised area is defined as any area not designated as a controlled area, but where occupational exposure conditions are kept under review, even though specific protective measures and safety provisions are not normally needed. For example, this could be a waiting area in the radiotherapy department.

3. Arrangements for restricting access to controlled areas. This is often done through badge access only doors or code-accessed doors. Radiation warning signs will also be displayed in these areas.

4. Arrangements for personal monitoring. This includes your personal dosimeter or dose badge. A dose badge will be given to you at the beginning of your placement, and you will need to swap it over as a measurement will be taken from the badge to ensure you have maintained a safe radiation exposure level. Your dose badge should be worn at chest height. You should not wash your dose badge, expose to direct sunlight or take on an airplane. If any of these things occur or you misplace your badge, you must let your university or practice educator know immediately. Please note that due to the very low risk of radiation exposure that therapeutic radiographers have, not all hospitals still use personal dosimeters.

Other items that are covered in the local rules include dose investigation levels, systems of work (procedures carried out to ensure dose will not be exceeded) and contingency plans for the department.

Radiotherapy bunker safety:

- On entry to the radiotherapy bunker, there are often barriers that create an interlock when activated. If disturbed when the radiation is on, the radiation beam will be cut off, this can be called a 'light curtain'. Other safety doors include physical barriers that need to be lifted, or machine-operated doors.

- The last person out button is a safety feature of the radiotherapy room, the button is pressed by the last radiographer to leave the room, signalling that everyone who needs to be out of the room is out.

- The emergency stop buttons are located throughout the bunker and in the control room. Learn where these buttons are located. If pressed, these buttons shut the machine down, therefore, stopping the radiation (if the machine is delivering radiation at the time).

Other radiation safety considerations on placement:

- You will hear radiographers refer to the 'pause and check'. There are often posters with the 'pause and check' checklists displayed in working areas. The 'have you paused and checked?' is a checklist followed by radiographers, before, during and after an exposure to ensure the radiation is delivered safely and correctly. The checklist includes patient identification, anatomy, user checks and consent.

Top tip: Look out for the checklist in your department.

- If you are pregnant or think you could be pregnant, it is important to inform your university as soon as possible. This is so that a risk assessment can be carried out, and the unborn child can be protected from any potential radiation exposures.
- If a patient has had a PET-CT scan, the staff will usually be made aware before their radiotherapy treatment. Hospitals have different ways of alerting staff they have recently had a PET-CT. If this is the case, pregnant staff/students should avoid treating that patient that day.

What you need to do:

- Read and follow the local rules in each area you work in
- Follow the procedures and work instructions
- Follow instructions from your supervising radiographer
- Wear your dose badge and change it when asked to do so
- Be aware of the emergency procedures for the area where you are working
- Report to your RPS/radiographer if you suspect a radiation incident has occurred

USEFUL DOCUMENTS AND REGULATIONS

Outlined below are some of the most important and relevant documents in radiotherapy practice (in the UK). As a student therapeutic radiographer, you will refer to them frequently throughout your training and eventually as a qualified radiographer. Important topics covered in these documents include image guidance, radiation safety, standard doses and fractionations and the code of conduct for radiographers. These documents are often referred to in exams and in interview situations and are reviewed regularly to reflect changes in practice.

The documents and regulations that will be discussed in this section are as follows:

1. Health and Care Professions Council (HCPC)
   i. Standards of Proficiency
   ii. Standards of Conduct, Performance and Ethics
   iii. Guidance on Conduct and Ethics for Students
2. The Royal College of Radiologists (RCR)
   i. Towards Safer Radiotherapy (2008)
   ii. Radiotherapy Dose Fractionation 4th Edition (2024)
   iii. On Target 2 (2021)
3. The Society of Radiographers (SoR)
   i. Code of Professional Conduct
4. Care Quality Commission (CQC)

Other important documents and regulations include The Ionising Radiation (Medical Exposure) Regulations (IR(ME)R) 2017 and amendment 2024, International Commission on Radiation Units and measurements (ICRU) 50, 62, 83 and the Ionising Radiation Regulations 2017, these are all discussed in other sections of the book.

HCPC

The Health and Care Professions Council are a regulator of health and care professions in the UK. The main role of the HCPC is to protect the public. They do this by regulating 15 health and care professions in the UK. Diagnostic and therapeutic radiographers are included in these professions as well as paramedics, physiotherapists and dieticians to name a few. It is required by law that if you are practicing in one of the regulated professions, you must be registered with the HCPC. When you qualify from an approved course, you can apply to be on the HCPC register. Once you have been accepted on the register, members of the public can look you up to check your registration. Your registration must be renewed every 2 years. In order to be a registrant, you must meet the HCPC standards for training, professional skills, behaviour and health. The HCPC takes action if professionals on the register do not meet the standards, and they investigate complaints against registrants. The standards outlined by the HCPC are summarised and explained below. It is important to visit the HCPC website for full information.

i. The Standards of Proficiency for Radiographers
These standards set clear expectations of registrants' knowledge and abilities when they start practicing. There are 15 standards and the HCPC website outlines each one in detail. The HCPC also have helpful resources on their website to help registrants meet and understand the standards (HCPC 2023).

ii. Standards of Conduct, Performance and Ethics
The role of these standards is to set out, in general terms, the expectations of registrants' behaviours and to outline what the public should expect from their health and care professional. These standards apply to learners on a HCPC approved programme and the HCPC has a document called 'Guidance on Conduct and Ethics for Students' on their website to support students understanding (HCPC 2024).

There are 10 Standards of Conduct Performance and Ethics (HCPC 2024).

1. Promote and protect the interests of service users and carers
2. Communicate appropriately and effectively
3. Work within the limits of your knowledge and skills
4. Delegate appropriately
5. Respect confidentiality
6. Manage risk
7. Report concerns about safety
8. Be open when things go wrong
9. Be honest and trustworthy
10. Keep records of your work

Some examples of how to meet these standards include using social media sensitively, maintaining appropriate professional boundaries and understanding the duty of candour.

The duty of candour is an obligation for health and care professionals to be open and honest with service users when things go wrong with their care (visit the HCPC, CQC or Government website for more information on the duty of candour).

It is important that you know, understand and meet these standards whilst you are a student and eventually a qualified member of staff.

RCR

The RCR are the leading professional membership body for clinical radiologists and clinical oncologists. Although the RCR is not the membership body for therapeutic radiographers, some of their documents and guidance apply to the radiography workforce.

i. Towards Safer Radiotherapy (The Royal College of Radiologists 2008)

   The aim of this report is to contribute to the delivery of an even safer radiotherapy service.

   The report documents what factors can contribute to a radiotherapy error, such as lack of training, fatigue and poor communication. Understanding these factors can help to reduce factors leading to errors and lead to a safer working environment. The report also defined a way to classify radiotherapy errors. These levels are as follows:

   Level 1: Reportable radiation incident

   Level 2: Non-reportable radiation incident

   Level 3: Minor radiation incident

   Level 4: Near miss

   Level 5: Other non-conformance

   Examples of errors that fall into each level can be found in the report.

   This is a very brief overview of the report. For further information, the report is available on the RCR website.

   Top tip: As a student, this document can help you to learn about why radiotherapy errors occur and how they can be classified. It is important to understand this in order to safely deliver radiation when you qualify.

ii. Radiotherapy Dose Fractionation 4th Edition (The Royal College of Radiologists 2024)

Compiled by a team of highly experienced clinical oncologists, using evidence-based practice, this document has helped to standardise radiotherapy prescriptions. The document is a guidance for clinical oncologists. It outlines the evidence base for each anatomical site and recommended treatment schedule. Please see Chapter 3: Oncological Management for an example dose and fractionation schedule.

Top tip: This is a very useful document when learning about dose and fractionation schedules. It can help you to understand the standard doses used for each anatomical site and the evidence behind it.

iii. On Target 2 (The Royal College of Radiologists 2021)

This report was commissioned to support the continued application of image guided radiotherapy (IGRT). The purpose of the report is to recommend best evidence-based practices for IGRT and recommend guidelines for implementing and optimising IGRT in clinical practice.

Top tip: As a student, this document can help you to understand what IGRT is and how/when it is used. It can also help you to understand the different pieces of equipment used in IGRT. The report outlines the verification process, geometrical uncertainties and adaptive radiotherapy. These are very important aspects of radiotherapy and knowledge of this report, and its content will help you in your studies.

SoR

The SoR is the trade union and UK professional body for the diagnostic imaging and radiotherapy workforce. This means that the society shape policy and standards, pioneer new ways of working and ensure safe and fair workplaces. You can join the SOR as a student radiographer for access to specialist webinars, lectures and conferences. The SOR also run competitions for students with monetary prizes for winners, visit their website to learn more.

One important document that the SoR have created is the 'Code of Professional Conduct' (The Society of Radiographer 2025). It sets out the SoR' expectations of the professional conduct of its members, which includes students, trainees and support workers. The importance of patient-centred care and patient safety is highlighted throughout. The document has four key sections;

Section 1: Relationships with patients and carers

Section 2: Patient safety

Section 3: Professionalism

Section 4: Professional practice

The SOR have published resources and guidance on their website to help members meet the code.

CQC

The CQC is the independent regulator of health and adult social care in England. That means that the CQC inspect all health and adult social care settings in England, for example, hospitals, care homes, dental practices and GP surgeries. The CQC asks five key questions of each provider:

Are they? 1. Safe, 2. Effective, 3. Caring, 4. Responsive to people's needs 5. Well-led

After an inspection the provider is given a rating of one of the following:

- Outstanding
- Good
- Requires improvement
- Inadequate

Top tip: It is helpful to know the CQC rating of your placement hospital/department and/or any hospitals you are applying for a job in. You can search the CQC website for the service or most hospitals display their rating and report on their website.

REFERENCES

Health and Care Professions Council. 2023. The standards of proficiency for radiographers. https://www.hcpc-uk.org/standards/standards-of-proficiency/radiographers/ (accessed 17 December 2025).

Health and Care Professions Council. 2024. Guidance on conduct and ethics for students. https://www.hcpc-uk.org/resources/guidance/guidance-on-conduct-and-ethics-for-students/ (accessed 17 December 2025).

Health and Care Professions Council. 2024. Standards of conduct, performance and ethics. https://www.hcpc-uk.org/standards/standards-of-conduct-performance-and-ethics/ (accessed 17 December 2025).

Legislation.gov.uk. 2017a. The ionising radiations regulations. https://www.legislation.gov.uk/uksi/2017/1075/contents (accessed 17 September 2025).

Legislation.gov.uk. 2017a. The ionising radiation (medical exposure) regulations 2017. https://www.legislation.gov.uk/uksi/2017/1322/contents (accessed 17 September 2025).

The Royal College of Radiologists. 2008. Towards safer radiotherapy. https://www.rcr.ac.uk/our-services/all-our-publications/clinical-oncology-publications/towards-safer-radiotherapy/ (accessed 23 September 2025).

The Royal College of Radiologists. 2021. On target 2: Updated guidance for image-guided radiotherapy.

The Royal College of Radiologists. 2024. *Radiotherapy dose fractionation*. 4th ed. https://www.rcr.ac.uk/our-services/all-our-publications/clinical-oncology-publications/radiotherapy-dose-fractionation-fourth-edition/ (accessed 17 May 2025).

The Society of Radiographer. 2025. Code of professional conduct. https://www.sor.org/learning-advice/professional-body-guidance-and-publications/documents-and-publications/policy-guidance-document-library/code-of-professional-conduct-2025 (accessed 15 December 2025).

CHAPTER 6

Practice Placement

INTRODUCTION TO PLACEMENT

Throughout this book, we have discussed the purpose of radiotherapy and its role within the cancer care pathway. This chapter will build on this knowledge by exploring the general principles of radiotherapy set-ups and your role in this process as a student therapeutic radiographer. It will support students to write useful objectives while out on clinical placement and guide them on coping strategies for managing difficult situations.

HOW TO WRITE SUCCESSFUL PLACEMENT OBJECTIVES

Setting placement objectives can sometimes feel overwhelming, especially at the beginning when you have so much to learn. This section aims to guide you on setting useful, achievable and suitable placement objectives.

Using SMART goals can help you to create successful goals. SMART is an acronym that uses five elements to help you set attainable objectives. Below are some radiotherapy-specific examples of how SMART objectives can be utilised.

Specific

Your goal should be clear and well-defined. It should answer what you want to accomplish, why it's important and how you plan to achieve it.

X Objective: To improve my confidence in patient set ups.

√ SMART Objective: To improve my confidence in setting up pelvis patients by using bed controls to align lasers to the patients' tattoos.

DOI: 10.1201/9781003639183-6

Measurable

Include something you can count or clearly observe.

X Objective: Set up patients under supervision to feel more confident.

√ SMART Objective: I will independently set up five pelvis patients, using the bed controls, under supervision by the end of the week.

Achievable

Ensure it's realistic given your level of training and the department's support.

X Objective: To lead all pelvis treatments by tomorrow.

√ SMART Objective: With support from my supervisor/mentor, I'll assist in daily setups until I'm ready to try it myself.

Relevant

Make sure the goal helps your learning as a therapeutic radiography student and is relevant to your year of study.

X Objective: I will spend time learning the reception computer software.

√ SMART Objective: Pelvis treatments and using bed controls safely are core competencies, so improving my setup skills will help meet my placement outcomes.

Time Focused

Set yourself achievable deadlines to stay on track.

X Objective: I will achieve this at some point.

√ SMART Objective: I will achieve this by the end of my 2 weeks on this machine.

A complete SMART objective: By the end of week 2 of placement on this machine, I will independently set up at least five pelvis patients using the bed controls to align the lasers to the patients' tattoos, with supervision, to improve my accuracy and confidence in treatment setup.

Try to create some example SMART objectives to help you meet your placement goals.

As well as writing successful placement objectives, another key element to completing successful placement objectives is knowing how best you learn. Take some time to think about this. The VARK questionnaire is a great way to help you figure this out. You can take this quiz online; it is a learning styles assessment tool. Knowing how best you learn can help you tailor how to engage in placement.

VARK will identify how best you learn out of the following:

V – Visual

Prefer to see information in maps, diagrams, charts, graphs and visual aids.

A – Aural (Auditory)

Learn best through listening – lectures, discussions, podcasts and spoken instructions.

R – Read/Write

Prefer information displayed as words – reading and writing, lists, notes, textbooks.

K – Kinesthetic

Learn best through hands-on experience – doing, touching, real-life examples, simulations.

You may be a multimodal learner which means that you learn best through a combination of the above. Once you understand what type of learner you are, you can apply this to placement for example: A read/write learner on placement might revise protocols by writing summaries and making flashcards. However, a kinesthetic learner might benefit most from hands-on setup practice and simulation.

SEEKING SUPPORT, RESILIENCE AND COPING STRATEGIES WHILST ON PLACEMENT

Therapeutic radiographers work with cancer patients daily; some of these patients can be very unwell, emotional or coming towards the end of their lives. This can have an emotional toll on the people caring for them, including you as student radiographers. It is important to acknowledge when you are affected by what you see on placement. Often, radiographers can seem unaffected by patients, but this is a learnt skill, and they have built their resilience during their career. Some of the things you see on placement you will be seeing for the first time, and it may come as a surprise to you what affects you and what doesn't. Some things to expect on placement that may affect you are listed below, consider how you may feel in these situations:

- Patients in large amounts of pain causing them to shout or scream
- Patients in emotional distress
- Patients deteriorating during their treatment
- Patients with open wounds that may smell
- Patients speaking about death and dying
- Patients who remind you of a relative

These are just some of the situations you may find yourself in, whilst on placement in a radiotherapy department. Reflecting on these situations is critical in order to build resilience. Chapter 7 outlines how to effectively reflect on situations that have challenged you. Reflecting on situations is one way to help you cope with them. Listed below are other ways to build resilience and support your emotional wellbeing during placement.

1. Accept that emotional reactions (stress, sadness, frustration) are normal.

   This is a new environment for you, accept these emotions are part of the learning process.

2. Seek support

   Speak to your mentor, practice educator or link tutor. Try not to internalise your emotions, and it is likely that one of these people will have experienced similar situations and emotions. You can also speak to your university wellbeing team about how you feel or ask about counselling services available.

3. Speak to friends, family and peers

   Figure out who your support system is, it might be housemates, family, friends or peers. Your peers on your course are an excellent support system as you will all be seeing similar situations at a similar time and learning together.

4. Find time for yourself
   Part of building resilience is to find time to relax and destress from the day. Build hobbies that help you to relax. Most universities have clubs that you can join.

5. Take time off if you need it
   If you are feeling stressed or overwhelmed, take some days off. Always communicate with your department and university.

6. Mindfulness and relaxation
   There are lots of free apps available that support these techniques.

7. Set boundaries
   Try not to take placement home. Find a way to 'switch off' after placement, for example maybe listening to music on your way home helps you to relax.

8. Stay active
   A healthy lifestyle can help manage stress.

Always remember why you chose this career, you want to help people, and this work makes a difference in people's lives. It's not always easy, and it's important to look after yourself or it can be a struggle to look after other people. Be prepared to see things that may affect you and put strategies in place. Building resilience takes time and seeking support is an important step in this process.

PRINCIPLES OF RADIOTHERAPY SET-UP

As a student radiographer, you will be expected to participate in the patient set-up process. To do this effectively, it is essential to understand the purpose behind set-ups. The primary goal is to ensure consistent and accurate patient positioning for each fraction. Accurate set-ups and reproducibility are vital for treatment precision and serve as a safety measure. Reading and confirming a set-up sheet, either in paper format or digitally, forms a key part of the 'pause and check' process prior to delivering treatment.

In most cases, a patient's set-up and associated equipment will remain consistent throughout the course of their radiotherapy. However, some variation can occur. Changes may be required due to factors such as swelling in the treatment area or weight loss. If anatomical changes are significant, the patient may require re-planning of their treatment to maintain accuracy.

Set-up Sheets

Set-up sheets are created at the patients' pre-treatment appointment, where a CT planning scan is taken and further detail can be added at dosimetry. The set-up sheet provides detailed information to the radiographers about the patient's identity, immobilisation method and positioning. In departments where permanent marks (PMs) are used as a reference point, this information will be included. Depending on the specific treatment technique and patient requirements, additional instructions may be noted. First, the patient must be as straight as possible. To do this, the radiographers check surface anatomy in line with the anterior laser, which is always centred on the patient.

Surface Anatomy

Surface anatomy refers to external body structures that represent internal anatomy. Once the patient is lying on the couch on their immobilisation, one radiographer will stand at the bottom of the treatment couch and check for straightness. If a patient is supine, they will check the anterior laser is running through mid-line structures. These include suprasternal notch (SSN), tip of xiphisternum (TOX), the root of penis (ROP) or vaginal cleft (VC).

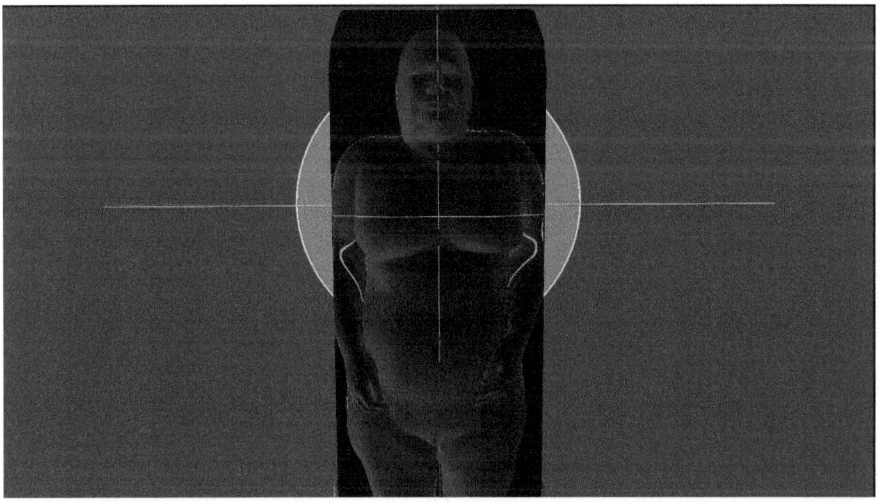

FIGURE 6.1 Mid-line anterior surface anatomy. (With permission from Vertual.)

If a patient is prone, the spine and intergluteal cleft (IC) can be used as surface anatomy landmarks. Please note that this is not an exhaustive list.

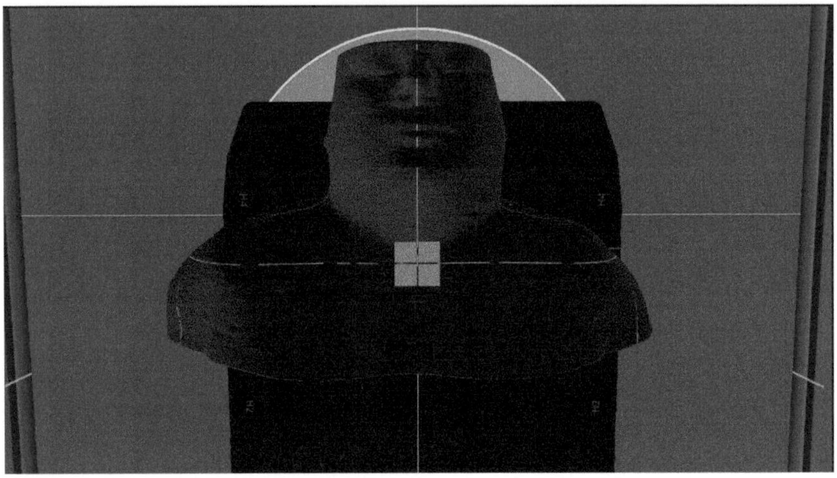

FIGURE 6.2 Anterior surface anatomy on a head and neck set-up. (With permission from Vertual.)

Some patients' natural position may not be through the structures identified above, and these details will be added to the set-up sheet to aid the radiographers at treatment set-up. As a student, one of the ways you can participate in a set-up is to straighten the patient's position prior to treatment/CT scanning.

Permanent Marks

PMs are given to the majority of radiotherapy patients to help support accurate and reproducible setups. These marks are given using a needle and ink, which is why they are permanent and are often referred to as tattoos. The most common number of PMs to be used is three, usually two are given laterally and one on the anterior of the patient (this is assuming the patient is in the supine position). PMs are given at the end of the pre-treatment appointment. These marks are drawn on with marker pen and only made permanent once the treatment position has been confirmed appropriate by the CT scan.

Firstly, the patient is positioned on the bed, the immobilisation is adjusted to fit them, and they are then straightened to sure they are lying on the couch as straight as possible. The surface anatomy used to straighten the patient is recorded on the setup sheet. Some patients naturally do not lie straight, if this is the case then their natural position is recorded to ensure that the therapeutic radiographers can reproduce this nonstandard setup.

The next step is to decide on the best position of the PMs. This is done using the lasers within the room, which are essentially used as a ruler to ensure all the PMs align. The CT couch can be moved, and the lasers can be manually adjusted to a position that is appropriate. When deciding on the location of these PMs the things to consider are:

- Ideally close to the treatment site
- On stable skin
- Not near any other marks such as moles or freckles that could be mistaken
- Away from any body hair

The location of the anterior PM can be decided by reviewing the skin around the treatment area to look for a suitable position, e.g. if the treatment site were in the pelvis such as the rectum, you would avoid being near pubic hair to ensure the PM is easily visible. An approximately suitable location would be 6cm superior to base of penis (BOP) in a male patient or 8cm superior from vaginal cleft (VC) in a female patient. If this position is appropriate, the lateral sides of the patient would be checked to see if this is a suitable position here also, if not the position can be adjusted superior/inferiorly until this is achieved. The next step would be to set the height of the lateral PMs, ideally this would be mid-separation. Again, if this is not quite suitable, you could adjust the anterior/posterior height of the bed or lasers to an appropriate position.

The three PMs should all align with each other in the superior/inferior direction. The anterior/posterior height should be set, and the anterior PM should be in midline; see image below of how this would look.

The therapeutic radiographers place radio-opaque markers on the pen marks they have placed to identify where the PMs will be. The radio-opaque markers are used to show up clearly on the CT scan. Once the therapeutic radiographers are confident, the position is appropriate and the CT scan takes place, they remove the radio opaque markers and replace by creating PMs. This means patients can wash off any pen marks, and the PMs will be left behind. It is important to note that these PMs are still very small and often patients are unable to see them.

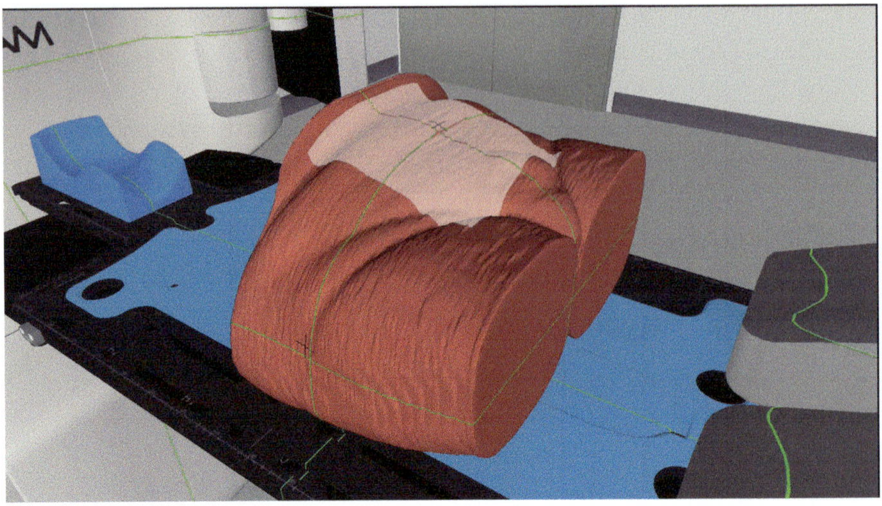

FIGURE 6.3 Example of a lateral and anterior PM outlined with a marker pen. (With permission from Vertual.)

However, this is why it's important for the position of the PMs to be recorded, to allow the therapeutic radiographers to locate them on the treatment units.

If once the CT scan is reviewed, it becomes evident that the three PMs no longer align, the dosimetry radiographers can provide alignments. A reason for this can be due to the patient moving slightly during the CT scan. An alignment can be made which will be measured and marked in relation to the PM, and this will be used for set-up purposes. This allows the therapeutic radiographers on the treatment unit to reproduce the position the patient was in when they received their CT scan, which is what was used for their radiotherapy treatment planning.

The position of the PMs, or alignments (if needed) become known as the set-up reference point. The radiotherapy plan is usually created so that the treatment is delivered to the centre of the planning target volume (PTV).

Once the setup is achieved, isocentric shifts will be applied (if given from dosimetry). These shifts have been calculated in dosimetry to move the position of the patient from the reference point to the isocentre. Therefore, the treatment will then be delivered to the isocentre.

Some clinical sites do not use PMs. These departments use surface guided radiotherapy (SGRT) which scans the patient's body with thermal cameras to ensure they are in the right place by mapping out the patient's skin surface.

Example Set-Up Sheet

Below is an example of a pelvis set-up sheet for a radical prostate patient. It is important to be aware that each anatomical site has its own set-up sheet layout and that not all clinical departments use the same format for set-up sheets. Therefore, you are encouraged to familiarise yourself with the specific documentation used within your clinical placement.

To solidify your learning, fill out the set-up sheet below as a case study. As always, remember to maintain patient confidentiality.

TABLE 6.1 Set-up Sheet

Patient details Name: Hospital number: Date of birth: address:	Patients picture Diagnosis: T2 N0 M0 prostate cancer Prescription: 60Gy 20# 4weeks
Patient orientation Head to gantry/Feet Supine/Post Hand on chest	Permanent marks Anterior: 6cm superior to root of penis Lateral: In line with anterior at a bed height of ....
Immobilisation Knee rest located at...... Ankle rest located at.......angle..... Head rest located at....... Mattress.... Bolus.....	Isocentre moves: Anterior: Posterior: Superior: Inferior: Left: Right:
Additional instructions:	Patient preparation: Bowel: Enema .... Bladder 350ml 1 hour before treatment
Pre-treatment signature: 1: 2:	

Additional Instructions

Additional instructions cover any considerations related to the patient's set-up and overall well-being. For instance, a patient with balance issues may need extra support ambulating, or they might use a stress ball for comfort during treatment. Some patients also require patient preparation, such as bladder filling, or taking certain medications at a certain time before treatment can be administered. These instructions can also document any modifications to the set-up or deviations from standard reference points, such as PM alignment or additional immobilisation devices. Depending on the treatment technique, further directions may be required, for example, implementing a deep inspiration breath hold (DIBH) for a left-sided breast patient.

Deep Inspiration Breath Hold (DIBH) and End Expiration Breath Hold (EEBH)

DIBH is a technique in which the natural process of respiration is manipulated. You may also hear it referred to as voluntary breath hold (VBH). When you breathe in (inhalation), your lungs fill and expand. This pushes the external tissue of the chest forward whilst the heart stays in its cavity. Exhalation contracts the muscles of the chest. This results in intrafractional target motion. By controlling the point at which inhalation ends, the radiographers can ensure that there is more space between the heart and the radiation. Studies have shown that this reduces lung, heart and pulmonary artery dose whilst maintaining homogeneity and tumour control without compromising CTV coverage. It is mainly used in the treatment of left-sided breast cancer but can also be used for centralised tumours and certain tumours of the abdomen. EEBH is the process in which a patient is asked to breathe out then pause breathing again until the radiotherapy has been administered. During exhalation and inhalation, the heart, lungs, liver and bowels move during each cycle of breath. By pausing this breathing cycle in the exhalation stage, it makes those organs less prone to movement. This is very beneficial when treating certain organs in the abdomen such as the liver. DIBH and EEBH are not the only instance where the bodies natural systems are used to minimise dose to OARs and maintain a homogenous dose throughout the CTV. As you can see from the above set-up sheet, one of the treatment requirements is bowel and bladder preparation.

Bowel and Bladder Preparation

This type of preparation is often used in the treatment of patients with cancer in their pelvis region. The premise of this preparation is to keep the treatment area stable, whilst offering maximum protection to the surrounding OAR. The type of preparation a patient undergoes differs for each anatomical site. Depending on the cancer site, patients will be asked to have either a full bladder and empty rectum or an empty bladder and empty rectum. As a student, you will become familiar with how to instruct a patient to complete these tasks. Below is a breakdown of these instructions, please note that these instructions are only a guide and can differ between clinical departments.

Example Instructions to Give a Patient Who Requires Bowel and Bladder Preparation

To achieve a full bladder and an empty rectum, drink 350ml of water, as fast as comfortably possible, 45 minutes before the appointment time. Do this after the enema has worked. Use an enema 1 hour before the treatment time. If it does not work, please report this to a member of staff. This preparation can be done at home or at the hospital at the patient's discretion.

It takes time for fluid to move around the body. The bladder is where waste liquid is stored. If you are dehydrated, this water will travel to other parts of the body resulting in a small bladder size. To help achieve a full bladder, it is important to stay hydrated throughout the day. Ideally, drink 2–3 litres of fluid a day. Please note that alcohol and caffeinated drinks lead to dehydration, so it is best to limit your intake of these.

An example timeline of this process is as follows: 9:00 am: Insert enema, wait for it to work and then empty rectum and bladder. 10 am: Drink 350ml water and hold bladder, 11 am: Treatment time.

Step-by-Step Guide for Setting up a Pelvis Patient

This step-by-step guide will take you through the entire process of a patient's daily set-up up to the point of leaving the treatment room. For this example, this guide will be based on radical prostate patients' daily treatment. However, the basics of this set-up can be transferred to any patient. Please note that different departments' set-ups may vary slightly.

In this section, you will also come across the terms 'control' and 'moving' radiographer. These are common terms used to describe the roles of each radiographer. The 'control', radiographer utilises the bed controls and the 'moving', radiographer is the person who will be administering a hands-on assist.

Throughout your time as a student, you will be involved in each step. It is important to only participate in the steps you are comfortable with and that each step is an opportunity to practice your communication skills. Remember, there will always be at least one qualified radiographer who will assist you.

Bringing a patient from the waiting room and into the treatment room:

1. Collect the patient from the waiting room.
2. In a quiet area, to maintain privacy, confirm that the patients' bladder feels comfortably full and their enema was successful.

3. Walk the patient to the changing room, if your clinical site has one, some patients may change in the treatment room. Patients will change into their treatment-specific clothing such as a gown or jogging bottoms. This makes the set-up process smoother for all parties and goes towards ensuring the patient feels comfortable and their dignity is maintained.
4. Once the patient has changed, walk them into the treatment room. The 3-point ID check will be conducted and details checked to make sure they match set-up sheet and any digital details.

Preparation of the treatment couch:

5. Remove previous immobilisation and clean the treatment couch.
6. Bring the gantry and collimators to the zero position.
7. Secure the patients' immobilisation onto the treatment couch and check the position next to the immobilisation sheet and digital information. Both treatment radiographers will check this.
8. Place a paper sheet over the immobilisation for infection control.
9. Bring the couch down to make it more accessible for the patient. Ensure bed is in the locked position to avoid falls.
10. Invite the patient to lay down. If appropriate, assist the patient with this step. Maintain appropriate manual handling.
11. Lay a modesty paper sheet over the patients' pelvis to maintain dignity.
12. If available, a bladder scanner will be used to check for bladder fullness. Communicate this process to the patient.

Moving the patient into the correct position:

13. Ensure the patient is comfortable and ask them to place their arms on their chest. This brings their arms away from the treatment area in-line with the ALARP principle.
14. Bring the treatment couch into the approximate position needed. Bed couch parameters will be captured within the first three days of a patient's treatment to assist with this.
15. Turn on the lasers and stand at the end of the bed to check for patient straightness. Use surface anatomy and anterior tattoo to assist with patient straightening.

16. If required, ask the patient to move into position. The 'moving' radiographer will ask the patient if it's ok for them to touch them and ask them to stay heavy on the couch; they will then assist the patient if necessary.

17. Locate PMs. You may see radiographers, 'mark', around the PM to make it easier to see. Ensure you check for any alignments from CT.

18. Set the bed height and long. The lateral will always be at 0, unless stated otherwise. Skip steps 19 and 20 if your department does not use PMs.

19. The radiographer on the 'control' side will move the bed into position where the lasers intersect in the middle of the lateral and anterior PMs.

20. The 'moving' radiographer will manipulate the patient's position, using small movements to achieve the same as the, 'control' radiographer. Throughout this process, the radiographers will communicate their PM position to each other.

21. Set a couch height (if used in your department, some departments use tattoo height).

22. Perform the isocentre moves and double check that the couch has gone in the correct direction.

23. Check the focus to skin distance (FSD).

24. Check rendered image.

IGRT preparation and leaving the treatment room:

25. If imaging, take out imaging panels. Depending on the IGRT, you may need to add filters or adjust the panel position.

26. Set the gantry position in preparation for imaging.

27. Let the patient know that you are leaving.

28. The last radiographer will press the last person out button. The last person out button makes a beeping noise to signal that everyone except the patient has left the room, and it is safe to administer treatment.

Outside the Room

Once step 28 is complete, the team will return to the control room, where the treatment consoles and monitoring cameras are located. From this room, radiographers can operate the equipment, take images and administer treatment with precision, all from a safe distance while ensuring patient safety. Many control rooms are also equipped with a tannoy system, allowing radiographers to communicate with patients throughout the procedure.

Treatment times can vary, typically lasting anywhere from a few minutes to around 30 minutes, depending on complexity. Once treatment concludes, the radiographers will re-enter the treatment room, inform the patient that the procedure is complete and encourage them to relax. The treatment couch will be lowered, and assistance will be provided if the patient needs help getting up.

As a student, it is always appreciated if you can assist by wiping down the equipment and removing any positioning aids in preparation for the next patient. The process then restarts from step 1.

IMAGE-GUIDED RADIOTHERAPY

Image-guided radiotherapy (IGRT) refers to any imaging used within the radiotherapy pathway. The aim of IGRT is to enable the radiotherapy treatment to be planned and delivered in the most conformal and accurate manner possible. IGRT is present in all aspects of the radiotherapy pathway from the images taken at pre-treatment to allow the consultant and radiographers to create the treatment plan, to the images taken on treatment to ensure accurate positioning. At pre-treatment, a patient is positioned on the immobilisation specific to their treatment site and a computed tomography (CT) scan is taken. The radiotherapy treatment is then planned on this CT scan dataset. At the treatment stage, the images from the pre-treatment CT scan become a reference image. This means that the images taken at treatment will be compared to the images taken at CT, and any differences at treatment need to be corrected for to ensure that the treatment position matches that at CT. Depending on the choice of reference image sent through to the treatment unit, can depend on how much detail is available to the therapeutic radiographers. From this comparison, the radiographers can make any adjustments that are required to ensure they are confidently treating the correct treatment site.

A reference image(s) can be:

1. 2-dimensional (2D), such as a digitally reconstructed radiograph (DRR).

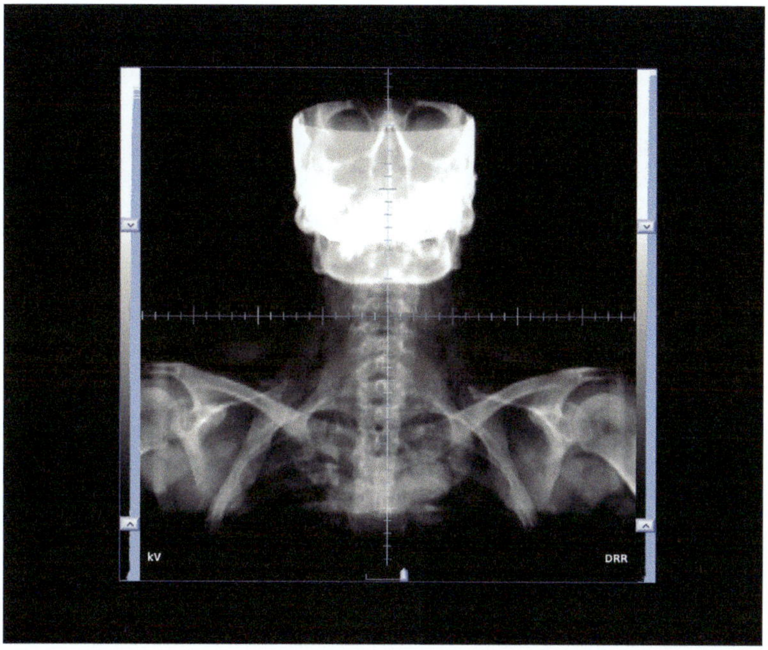

FIGURE 6.4 2D DRR image. (With permission from Vertual.)

2. 3-dimensional (3D) imaging, a section of the CT scan dataset near the treatment area.

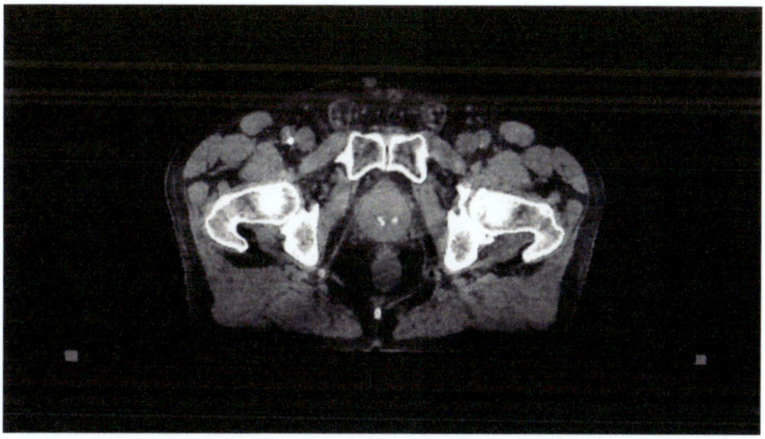

FIGURE 6.5 A slice from a 3D CBCT. (With permission from Vertual.)

3. 4-dimensional (4D) imaging, a section of the CT scan with added patient movement such as breathing.

The treatment technique chosen greatly influences the type of IGRT the therapeutic radiographer will use. The more advanced the treatment technique, the more likely a more advanced form of imaging will be used.

Most modalities that deliver external beam radiotherapy have integrated imaging, therefore providing IGRT. The most common imaging modalities are megavoltage (MV), kilovoltage (kV) or CBCT using kV or MV. Each imaging modality has its own advantages and disadvantages, which must be considered when deciding on the appropriate IGRT.

TABLE 6.2 Imaging Modalities

Modality	**Advantages**	**Disadvantages**
Megavoltage (MV)	• defines bony anatomy well.	• More dose compared to kV imaging.
Kilovoltage (kV)	• Defines soft tissue anatomy well. • Lower dose compared to MV images.	• Reduce definition of bony tissue compared to MV imaging. • Unable to penetrate metallic structures e.g. an artificial hip.

MV Imaging

2D MV images are produced using electronic portal imaging (EPI), this requires a flat panel to be positioned opposite the treatment beam producing a digital image.

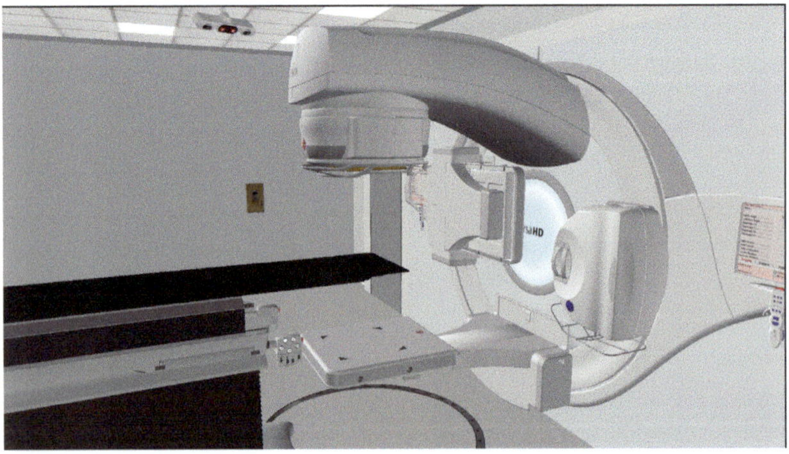

FIGURE 6.6 MV panel on a linear accelerator. (With permission from Vertual.)

This imaging technique can use the treatment beam itself to take an image or a small beam can be delivered before or after the treatment to create an image. An advantage of using the treatment beam itself to take an image is no additional concomitant dose is delivered to the patient, as the monitor units (MU) from the treatment field itself can be used. However, this technique only allows for a 2D image to be created. This can be supported by taking orthogonal images such as an image at 0 degrees and one at 90 degrees.

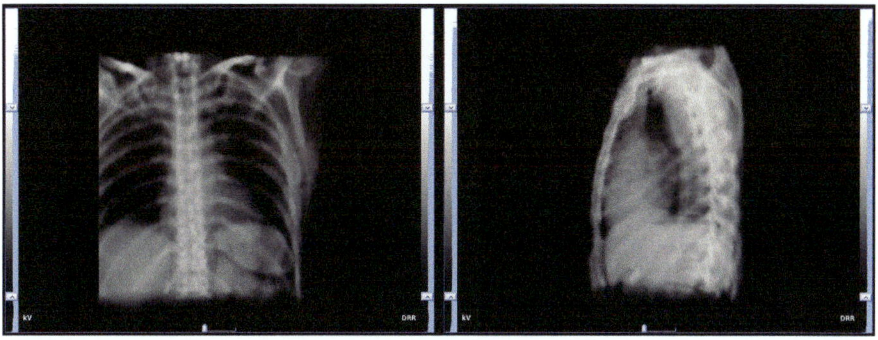

FIGURE 6.7 A 2D Orthogonal pair. (With permission from Vertual.)

Therefore, giving data in the longitudinal, lateral and vertical directions. MV images can be taken of the treatment field itself which is known as a single exposure or the jaws can be opened to a bigger field size to allow for more anatomy to be visible to aid treatment accuracy. This is known as a double exposure. Other ways that MV images can be utilised is by having a separate imaging field, which allows for a field size to be chosen that incorporates all the relevant anatomy required to make an accurate image comparison.

kV Imaging

For kV images, these are taken before the treatment where an additional imaging field is created at either the treatment angle or a more useful angle if necessary. Therefore, the imaging field is separate to the treatment field and the imaging field size can be different to the treatment field size, if again thought to be beneficial. kV images can also be created at orthogonal angles such as 0 degrees and one at 90 degrees to give data in the longitudinal, lateral and vertical directions.

Advancements on imaging are the use of cone beam computed tomography (CBCT) imaging, which most current linacs are integrated with. A kV arm is integrated into the linac design to allow for the CBCT images to be taken.

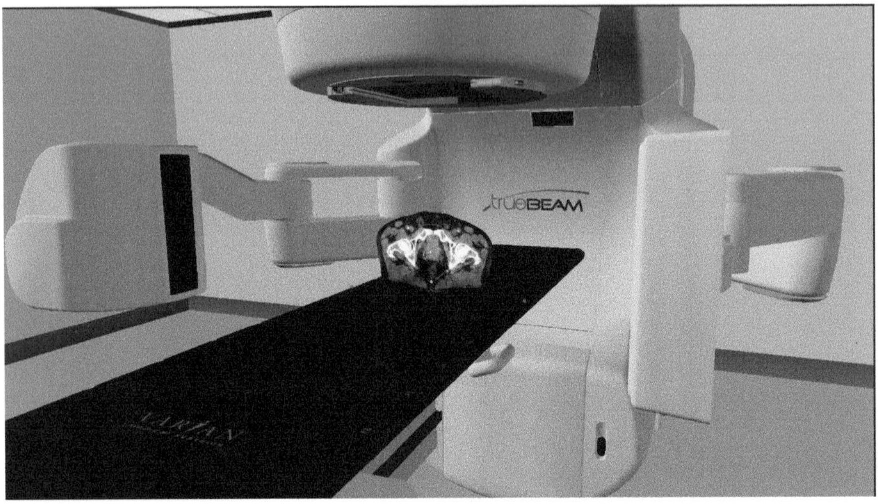

FIGURE 6.8 A CBCT image of a prostate. (With permission from Vertual.)

This allows for volumetric imaging of an area of interest, giving more data which can be compared alongside the pre-treatment CT reference image. This type of imaging can use either MV or kV, but must be taken separately to the treatment, usually before the treatment is delivered.

Depending on the treatment being delivered and on departmental protocol, it can detail how often IGRT is carried out and what type of IGRT is taken. Imaging can be taken before the treatment is delivered, during the treatment delivery and after treatment delivery. Imaging can be taken at every fraction, or it can be less often such as weekly.

CHAPTER 7

Reflective Practice

INTRODUCTION TO REFLECTIVE PRACTICE

This chapter will focus on reflective practice and the importance of this as a student therapeutic radiographer. Reflective practice can be linked to personal and professional development; therefore, it will be evident within your university course in multiple ways. The focus here will be reflective practice to develop you as a professional student therapeutic radiographer.

WHAT IS REFLECTIVE PRACTICE

Reflective practice allows you to think back to a situation retrospectively, identifying how you and others dealt with the situation at the time. This can be completed independently and/or as a group. A health care setting can move at a very fast pace, and it can be difficult to make the right decision in the moment. This is where reflective practice is most beneficial. Taking time to think back to a situation and analysing your own and others actions is how you can decide what was carried out well and what improvements are needed to ensure good practice. Reflective practice allows a situation to be reviewed in a calm environment with no time pressure.

As reflective practice can be so beneficial, you may experience this in many ways whilst out in the clinical department. It could be that a situation occurs in the radiotherapy department, and you are asked as

DOI: 10.1201/9781003639183-7

part of a team to reflect on the situation, critically identifying areas for improvement and identifying good practice. You may also be required to complete individual reflections on topics of your choice that you feel have impacted on your professional development as a student therapeutic radiographer.

However, it is important to note that reflective practice can be quite a difficult skill to learn. It is easy to describe a situation that occurred and think back at how you dealt with that situation; however, this is not reflective practice. For reflective practice to be carried out, you need to deeply examine the situation and scrutinise your practice. This will enable you to identify good practice, with the main purpose of maintaining this, or it will help to identify areas of improvement which can be implemented to avoid the situation occurring again. This is how we learn and share lessons of best practice. Reflective practice can be carried out on situations that went well or situations that were challenging; this is down to personal preference and what supports you as an individual and your professional development.

REFLECTIVE MODELS

To support reflective practice, there are models of reflection which can be drawn upon. These models vary slightly, but all have the same aim: to encourage the process of critical reflection and try to reduce the descriptive element of reflection. The more you engage in reflective practice, the easier it can become. If you are new to reflective practice, a reflective model which can be quite beneficial is Gibbs reflective cycle (Gibbs 1988).

This model allows for more description with 6 key questions and does include some analysis. Whereas Rolfe's model of critical reflection, which is a very simple model, asks three key questions: "What, So What, What Next" (Rolfe *et al* 2001). Therefore, focusing on analysis of a situation and not on description, which is why it is often suited to a more experienced reflective practitioner. As mentioned previously, there are lots of different reflective models to support reflective practice which can be used, or if you are confident at reflective practice, you do not need to use a model at all.

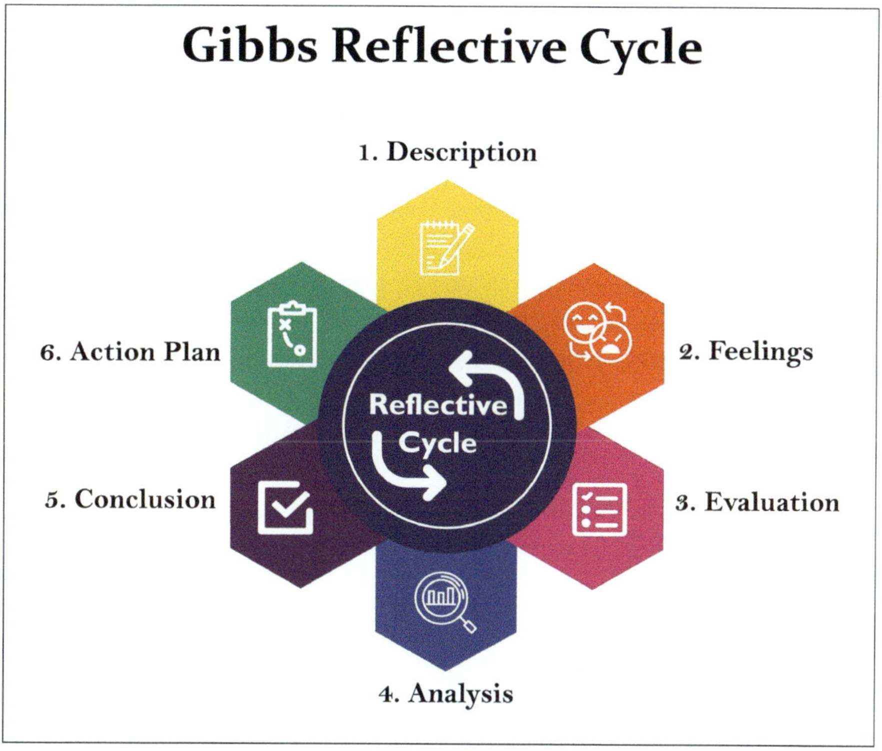

FIGURE 7.1 Gibbs reflective cycle.

REGULATORY BODY REQUIREMENTS

The HCPC have four standards relating to reflective practice and continuing professional development (CPD), located within the standards of proficiency and standards of conduct, performance and ethics.

Standards of proficiency:

- Keep their skills and knowledge up to date and understand the importance of CPD throughout their career (1.3).
- Understand the value of reflective practice and the need to record the outcome of such reflection to support continuous improvement (10.1).
- Recognise the value of multi-disciplinary reviews, case conferences and other methods of review (10.2).

Standards of conduct, performance and ethics:

- You must keep your knowledge and skills up to date and relevant to your scope of practice through CPD (3.4)

Reflective practice can help support the meeting of these four standards; therefore, it is important that you develop the skills of reflecting as a student therapeutic radiographer. Reflective practice will support your professional development, which you can then continue to build upon as a qualified therapeutic radiographer. Reflective practice is a critical part of our professional development and employers support the need for this. This is especially important if you are reflecting on situations/activities that you feel did not go well, to avoid this happening again. To enable this when reflecting on practice it is essential you are open, honest and critical.

The HCPC provide helpful resources to support reflective practice on their website (HCPC 2021).

AN EXAMPLE OF REFLECTIVE PRACTICE USING THE GIBBS' REFLECTIVE CYCLE

1. **Description**
   The treatment machine I was working on was running 45 minutes behind. I was asked to write the time delay on the whiteboard for patients to be able to see. A patient became agitated and verbally aggressive towards me when they noticed the time delay for their treatment.

2. **Feelings**
   I felt anxious and didn't know what to do. I was worried that the patient might become physically aggressive towards me. I also felt embarrassed as there were other patients in the waiting room.

3. **Evaluation**
   I stayed calm in the situation, but I was unsure of how to apply my de-escalation training, so a senior member of the team came over and helped me.

4. **Analysis**
   This situation highlighted a gap between my theoretical understanding and how to apply it in a practical setting. More practice in these types of situations would help me.

5. **Conclusion**
   I could have anticipated that the patients would have been upset by this delay and asked for a senior member of the team to support me from the beginning.

6. **Action plan**
   I plan to continue practicing de-escalation strategies using virtual software and to practice with my peers. I will also ask for feedback from the team on how to manage similar situations in the future.

The HCPC website includes templates to support reflection and case studies, which can be used to help you practice reflecting.

REFERENCES

Gibbs, G. 1988. *Learning by doing: A guide to teaching and learning methods.* Oxford: Further Education Unit, Oxford Brookes University.

Health and Care Professions Council. 2021. Reflective practice. https://www.hcpc-uk.org/standards/meeting-our-standards/reflective-practice/ (accessed 23 May 2025).

Rolfe, G., Freshwater, D., Jasper, M. 2001. *Critical reflection in nursing and the helping professions: A user's guide.* Basingstoke: Palgrave Macmillan.

CHAPTER 8

Academic Support

INTRODUCTION TO ACADEMIC SUPPORT

As a therapeutic radiography student, you will often be required to undertake academic aspects during your studies. It will be expected of you to use literature in your work and be able to reference this literature. Understanding the need for referencing and how to do it correctly is a vital part of academic studies. This chapter outlines the type of evidence and gives examples of how to reference to help you be successful in the academic element for your studies.

EVIDENCE-BASED PRACTICE

Evidence-based practice is an approach to clinical practice that uses evidence from research, clinical expertise and lived experiences to enable healthcare professionals to make the best decisions. This approach is widely used across healthcare and can be applied directly to radiotherapy.

Radiotherapy treatment prescriptions and side effect management are all developed through evidence-based practice, and recommendations are provided to ensure gold standard treatment is delivered. NICE guidance provides recommendations in the health and social care sector based on evidence-based practice from a variety of sources such as professionals and service users, with the overall aim to improve healthcare.

Research can be classified into different levels. Evidence-based practice relies on the best available evidence, which comes from research. Research can take place through a variety of ways; however, research that has been

DOI: 10.1201/9781003639183-8

peer-reviewed is considered to be the most valuable. Other ways that evidence can be collected are through work-based trials, clinical trials, professional experience and feedback. The use of evidence-based practice enables patients to make informed decisions, knowing that professionals have reviewed all of the available evidence.

PRIMARY AND SECONDARY RESEARCH

When reviewing research, it is important to understand whether it is primary or secondary research. Primary research refers to research where the data has been collected from the original source, first-hand data collection by the researcher, such as through methods of interviews, focus groups, observations or surveys. Whereas secondary research is where data that has already been collected, and then is reviewed, interpreted and/or analysed by others. This can then be presented in articles, books, reports, etc. Therefore, primary research is considered to be more valuable as it consists of the raw data. It allows for questions to be asked that may have never been asked before. Whereas secondary research relies on the researcher analysing another researcher's data. However, both types of research have a place, although secondary research is interpreting someone else's data, it still allows for the researcher to gather insight into previous research. The most common method of secondary research is a systematic review, where researchers can review existing literature and studies through systematic methods appraising studies to summarise their findings. Although this might sound daunting, there is a lot of guidance on how to implement systematic methods to review literature. The main thing to note at this stage is the difference between the two research types.

ACADEMIC REFERENCING

Academic referencing is important when completing academic work as it helps to avoid plagiarism and gives credit to the source information and where it came from, strengthens academic arguments and allows readers to verify the sources.

It is important to note that different universities use different referencing styles. It is important to know which style your university uses. This section will give information on referencing which can be applied to all styles.

Some key referencing concepts are:

In-text citations

In-text citations are when the reference is placed in the text to alert the reader to the source that has informed the writing. The citation is often a shortened version of the full reference. The in-text citation should have a corresponding full reference in the reference list. In-text citations should be used in essays, presentations and posters.

Example: Paraphrasing – putting the sources work into your own words

In-text citation (Author, year)

The study showed a significant improvement in children's health.... (Davis, 2023).

For two authors, the citation would be (Davis and Wilson, 2023) and for more than two authors you can use et al. (Davis *et al.*, 2023)

Example: Direct quotation – directly taking the words from the source

In-text citation Author (year, page)

Davis (2023, p.45) argues that 'children's health can be improved by....'

Example: Starting a sentence with a citation

Davis *et al.* (2023) concluded that a significant improvement in children's health......

Reference list

The reference list comes at the end of the academic work, and full details of the in-text citations are given. The way in which the references are laid out will depend on which style you are using and what type of work you are citing. Typically, reference lists are alphabetised although some referencing styles follow a numbering pattern.

Example: How a journal article may be referenced (Harvard referencing)

Author's surname, Initials. (Year of publication) 'Title of the article', Title of the Journal, volume number (issue number), page range of the article.

Example: How an online video may be referenced (Harvard referencing)

Author's surname, Initials or username or Organisation or Pseudonym (year created/uploaded) Title of video. Available at: www.webaddress.com (Accessed day month year)

Bibliography

A bibliography can be used to demonstrate broader reading than a reference list and includes uncited works.

Common mistakes to avoid:

- Using bullet points and not alphabetising
- Mixing styles
- Inconsistent punctuation/capitalisation
- Citing without including in reference list (and vice versa)
- Attaching URLS that are incorrect

Practical tips:

- Keep a running list of sources while writing
- Use reference management tools such as RefWorks
- Double-check details (author names, dates, URLs)
- Read your institution's specific style guide, small differences matter

ARTIFICIAL INTELLIGENCE (AI) IN ACADEMICS

Artificial intelligence (AI) is very popular right now, and as a student, it can be a very helpful tool. However, AI needs to be used with caution and understanding. This section aims to help you understand what AI is and things to consider when using it. Please note that it is important to check your institution's guidance of usage on AI.

What Is AI?

AI is a broad field of creating machines that simulate human intelligence. It is the development of computer systems to carry out tasks that previously required human intelligence, such as reasoning, learning, problem solving and decision-making.

Examples: Amazon recommendations, spam filters, Google Maps, facial recognition, etc.

Generative AI (GenAI) is a branch of AI that creates new content – such as text, images, video or audio – based on patterns it has learned from large datasets.

Examples: ChatGPT and Copilot

Benefits of AI

- To get a general introduction to a subject area
- To summarise a complex topic, or to break down a complex task
- To generate some initial ideas, for instance around the structure of an essay
- To help you spellcheck
- As a study tool to help generate practice questions
- As an organisation tool to help you organise your study rota
- To support accessibility

Limitations of AI

Copying the words, arguments or ideas from an AI tool without proper citation and referencing (in other words, asking an AI tool to write an answer to your assignment question) would count as plagiarism.

Furthermore, AI tools can sometimes provide incorrect or entirely fabricated information in response to a question or topic that they don't know much about. AI tools often create 'hallucinated' references that do not actually exist.

They can also be biased towards a certain view of the world depending on the content of the data they have been trained with. AI also tends to lean towards the strengths or positives rather than limitations and negatives. AI tools are not very good at critical thinking, reflections or identifying personal experiences, which may be asked of you in a piece of academic work.

Therefore, it is important to always check AI-generated content against your lecture materials, reading lists and credible academic sources. Use these tools to enhance your understanding, not as a substitute for critical thinking.

CHAPTER 9

Study Tips and Tricks

INTRODUCTION TO STUDY TIPS AND TRICKS

Therapeutic radiography is a broad subject with many different aspects to learn. This, combined with a clinical placement, can make keeping on top of coursework and studying challenging. The amount of information to remember can feel overwhelming. This book has introduced you to many topics including oncological management, physics and clinical placement. Within each of these subjects, there are several key factors to remember and build upon. This section will provide tips and strategies to help you monitor your learning and reduce some of the stress and pressure associated with completing assignments and exams.

REVISION TIPS

Understanding your learning style can give you valuable insight into the most effective ways to study. Regardless of your learning style, it is always important to attend all lectures and clinical practice.

Non-learner Specific Advice

Regardless of your learning style, consistency is one of the most effective approaches to studying. Creating a study schedule that fits your lifestyle is an important tool for success. Many students assume that the number of hours spent studying is what matters most, but pushing yourself to study for long periods can increase stress and reduce information retention. A helpful strategy is to set aside a manageable and realistic amount of time after lectures to review and expand on your notes. Writing your own questions not

DOI: 10.1201/9781003639183-9

only clarifies your understanding but also helps you build a 'question bank' to prepare for exams, which can be even more effective when shared within a study group. Finally, taking time to reflect on your practical experiences can deepen your learning and strengthen your overall understanding.

Top tip: It can feel intimidating to ask lecturers questions; however, it is important to remember that you will not be judged for asking a question. Everyone learns differently; you may require a different explanation or tool to help you understand a concept.

EXAM AND ESSAY STYLE QUESTIONS

The meaning of essay or exam questions can sometimes be difficult to interpret. Often, it is not the topic itself that can cause confusion, but the precise wording of the question. Even a single word can significantly alter what is being asked and how the answer should be structured. Table 9.1 gives examples of commonly used terms in a university setting with example questions that you can answer in your own time to solidify your learning.

TABLE 9.1 Examples of Commonly Used Terms in a University Setting

Words	Explanation	Example Questions
List	Write bullet points in relation to the subject. No elaboration is required.	List 3 side effects of radical radiotherapy to the prostate.
Discuss	Talk about the subject matter: what, why and how. If possible, also add, alternative options and evidence from literature to support your answer.	Discuss the difference between VMAT and IMRT.
Define	To give a short definition that's concise and to the point.	Define the acronym IMRT.
Describe	A definition with more depth. Essentially, it is a clear description of that topic.	Describe the acronym IMRT.
Justify	An answer that includes clear reasons and evidence to support your answer.	Justify the use of IMRT in the treatment of radical prostate cancer.
Evaluate	A concise critique of the subject matter with evidence to support your discussion.	Evaluate the use of IMRT in the treatment of radical prostate cancer.
Identify	Find the key facts. This could be in the form of labelling a diagram or a focus on a key area of a subject matter.	Identify the main characteristics of IMRT.
Outline	Give a brief and concise summary of the subject.	Outline the use of IMRT.

Reading a Question

Now that you are familiar with keywords and their meaning, we will now review how to read and interpret a question. Mis-reading a question is one of the top reasons that students lose marks in assignments and exams. In any circumstance, always read a question at least three times. If possible, read the question aloud. This repetition will give you time to absorb and comprehend the meaning of the question. You can then underline or highlight key words in the question; this can help you to break down the question. It is also helpful to make a note of what you will include in your answer to help focus your answer.

Example question 1:

Describe the radiotherapy technique utilised for the treatment of T2N0M0 prostate cancer.

Your answer should include:

- Patient positioning
- Patient dose and fractionation
- Treatment technique

Breaking down the question: Key words are in bold, and notes are in italics

Describe *(A definition with a more depth)* the radiotherapy technique utilised for the treatment of **T2N0M0 prostate** *(This indicates that the treatment will be of a radical intent. The question does not ask me to describe the TNM system)* cancer.

Your answer should include:

- **Patient positioning:** *How is the patient positioned on the treatment couch? What kind of immobilisation does this patient require?*
- **Patient dose and fractionation:** *What is a full radical prostate radiotherapy prescription? A full prescription includes Gy, # and timeline.*
- **Treatment technique:** *IMRT is a good option for this patient. What is IMRT? Why is it beneficial? What type of IGRT would be suitable and why?*

Example question 2 with breakdown:

Patient-led and disease-led factors are reviewed before deciding on the type of **oncological management** a patient could receive. *Question is not asking me to describe oncological management.*

Identify *(key facts, no elaboration required)* the **patient-led** and **disease-led** factors that will be reviewed. *Can make a list of patient and disease-led factors.*

Answering a Question

Once you have broken down the question and are ready to answer, there are a few rules you should follow.

- Avoid using colloquial (slang) language. This can lead to misinterpretation and is not considered professional.
- Where possible use scientific language. For example, using the word erythema instead of sore skin.
- Always use the full name before writing an acronym. For example, MLC should first be written as multileaf collimators. While acronyms can be helpful, their overuse or use without explanation can make answers unclear.
- Only answer what the question has asked you. For example, refer back to example question 1. The question gives you a diagnosis; however, it has not asked you to describe the staging system. Always take your time to fully understand what is being asked before beginning your answer.
- Ensure completeness of responses: Make sure that each part of the question is addressed thoroughly. Pay careful attention to the words being used: discuss, define, identify.
- Avoiding Repetition: Aim to present a broad and varied range of information to support your answers.

All of the information above can be applied to any assignment or exam. However, there are some specific tips to tricks that can assist you in individual situations which are document below.

EXAM TIPS

In an exam, always take note of the marks allocated to each question, as this will guide the level of detail required in your answer. For example, if a question is worth five marks and asks you to list five aetiological factors of lung cancer, providing ten answers will not earn you extra marks, as the examiner will only consider the first five correct responses.

It is equally important to clearly indicate which question you are answering by writing the correct question number next to your response. Many students lose marks simply because they fail to number their answers correctly or match them to the right question.

A useful strategy is to immediately jot down any key information you recall on a spare sheet of paper before you begin the exam. Invigilators can provide this. This creates a 'library' of information that you can refer back to and expand on throughout the paper.

Although exams can feel stressful and the urge to dive straight in may be strong, it is best to take your time when reading the questions. If possible, read through the entire paper at least twice, then begin with the question you feel most confident answering.

To support your performance, avoid excessive caffeine beforehand, ensure you get a good night's sleep, stay hydrated and take a short walk to clear your mind before the exam. Listening to calming music, a favourite song, or a podcast can also help reduce anxiety. Your university may also offer courses and advice on how to decrease exam anxiety which you may find useful.

Finally, if available, make the most of mock examinations. These provide valuable experience in the exam setting and help familiarise you with expectations, which can significantly boost your confidence on the day.

COURSEWORK TIPS

Timing is crucial, avoid leaving your coursework until the last minute. Start by making a clear plan that allows you enough time to fully understand the question, review the brief (if provided), examine the marking criteria, and gather your research materials. Once these steps are complete, you can begin drafting your coursework with confidence.

After carefully reading and breaking down the question, your next step is to collect relevant information. This can come from lectures, personal notes, textbooks, journal articles and recommended reading lists.

While it may be tempting to follow interesting tangents in your research, it is important to remain focused on the specific question being asked. Keep your work concise, relevant, and directly aligned with the assessment criteria.

When gathering evidence, always prioritise credible, up-to-date and relevant sources. This is especially important in scientific subjects where knowledge and practice evolve quickly. Peer-reviewed journal articles and recent publications will strengthen your arguments and demonstrate critical engagement with current research. Remember to always use your university guidelines for referencing. Referencing is an extremely important part to any coursework as it lends credibility to your writing. Remember to keep any sensitive information such as a patient name confidential.

To ensure your writing is easy to follow create a rough structure or outline of what you would like to include. As you research, categorise information under potential headings or themes. This not only helps you stay organised but also ensures your argument flows logically when you begin writing. By managing your time effectively and keeping your research focused, you can produce coursework that is both well-structured and evidence based.

Always read around your subject to the extent that you would feel comfortable explaining your writing and answering any questions.

Layout of Coursework

With any coursework, whether it be an essay or a presentation, how you layout information can make a big difference. A basic essay layout is as follows:

- Title: The assignment question
- Introduction: Introduce your subject to the reader. What are your aims? What topic will be discussed? This can usually be achieved through 1–2 paragraphs. Remember to write as if the person reading this has no prior knowledge of the subject. Try not to assume that something is 'obvious', if you do not write it down, it cannot be marked.
- Body of work: This is where you will expand on and discuss the aims that you pointed out in your introduction. This will be the main section of your work.

- Conclusion: This is a summary of the key points that were stated throughout your essay. It should link up nicely with the essay title and aims. Remember a conclusion is a great opportunity for reflection. However, a conclusion should not include any new information.
- Reference list.

Top tip: Unless indicated, such as in a reflective piece, do not write in the first person.

Presentation Tips

Effective presentation skills are an essential part of academic and professional development. The ability to communicate ideas clearly, confidently and persuasively not only enhances your coursework but also prepares you for future clinical and professional settings.

When planning a presentation, always consider the needs of your audience. Ensure that your slides are easy to read, with a clear font and text size, and that any visuals, such as images, charts or diagrams are relevant to your topic. Keep the layout simple yet engaging, using colour and visuals to support your points without overcrowding the slide.

Equally important is your verbal delivery. Speak clearly and at a steady pace, and practice delivering your presentation out loud beforehand. This will help you become comfortable with the flow of your content and allow you to rehearse any terms or technical words that may be difficult to pronounce. Aim to reduce filler words (such as 'umm' or 'like') and maintain professional, scientific language, while also using open body language to keep your audience engaged.

If you are using technology during your presentation, practice with the device and software several times in advance. Familiarity with the setup will help you feel more confident and prevent avoidable technical difficulties, reducing stress on the day.

Poster Tips

Posters can appear deceptively simple but creating an effective academic or scientific poster requires careful planning and design. One of the main challenges lies in filtering through large amounts of information and presenting it in a way that is not only visually engaging but also clear, logical, and easy for the audience to follow. A strong poster should balance concise text with impactful visuals, highlighting key points without overwhelming the viewer. Below are some top tips in how to create a poster:

- Keep it concise: Focus on key points and avoid unnecessary detail.
- Make it readable: Choose legible fonts, use large enough text and ensure there is good contrast between background and text.
- Visual impact: Use graphs, diagrams and images to complement your text, but avoid overcrowding.
- Consistent design: Stick to 2–3 main colours and a simple layout.

CONVERSATION STARTERS

As previously discussed, communication is a vital aspect of being a student. During your studies, you may encounter situations where communicating feels uncomfortable, intimidating, or confusing. Below are some common examples of such scenarios, along with straightforward strategies you can practice building confidence and strengthen your communication skills. The scripts are kept simple and direct to minimise any confusion for both parties. It is also important to remember that listening is just as essential as speaking.

In all patient and staff scenarios, if it is your first time meeting them, it is best practice to always introduce yourself first following the format in Chapter 6. Remember, if you ever feel uncomfortable or upset in any scenario, you can always defer to a member of staff that you trust. Alternatively, you can also contact your practice educator, fellow students or a tutor for advice.

Patient Scenarios

Asking a Patient a Question

'Hello (patients name), I would like to ask you a question'

Keep the question simple and to the point to avoid confusion. If the patient declines to answer or there is a barrier to communication defer to a qualified member of staff.

This line can be adapted depending on the scenario. A simple gateway to help assess someone's overall wellbeing is, 'Hello (patients name), how are you feeling today'.

Asking a Patient to Move on the Bed

In this scenario, it is important that you do not use confusing jargon such as superior and inferior. Instead, use phrases such as move towards your feet or move up the bed.

'To ensure you are in the correct position (patients name) could you please move to the (insert direction) very slightly'.

If you will be touching a patient for any reason always let them know what you are doing by using the following line, 'Hello (patients name) to get you into the correct position is it OK if I touch your (insert anatomy e.g. arm/chest/pelvis)', 'Please lie heavy on the bed and do not try to help me as I only need to move you a small amount'. If you have cold hands, let the patient know before you touch them to avoid any shocks.

A Distressed Person

This script can be adapted to suit anyone; cancer does not just affect the patients but the people surrounding them. Your fellow students may also feel upset at certain points.

'Hello (insert name) I can see that you are feeling (insert emotion)'

You can then pass this observation onto a member of staff by telling the person, 'With your permission, can I share that you are feeling (insert emotion) with a staff member who may be able to assist you'.

Alternatively, if you feel comfortable, you can ask, 'would you like to share what made you feel this way?', 'If so I am here to listen'. It is best to guide the person to a quiet and private place, always let a staff member know if you will be leaving the set and why.

You may come across situations where someone will share a personal thought or situation with you such as 'I am scared' or 'I am afraid I might die'. These situations can vary and as such there is no script as it is a continuous learning process. However, there are some key sentences that can be effective.

'I'm very sorry to hear this'

'Thank you for sharing that information with me'

'Is there any specific help or support you need'

'With your permission may I communicate this with the members of the team that are best placed to help you?'

Top tip: The above sentences can also be used if someone asks you a question you are not qualified to answer.

First Day Chats

First day chats are the perfect opportunity to practice all of your communication skills.

First day chats are an integral part of a patient's radiotherapy treatment. This is where you will determine that the patient understands what the process of radiotherapy is, the risks and benefits and if they still consent to treatment. Remember, consent is an ongoing process and can be withdrawn at any time for any reason. Below is a standard script that you can use to build upon for any patient.

- 'Hello, my name is….'
- 'I am a student radiographer'
- 'I will be assisting the radiographers with your first day chat today'
- 'Before we begin, I have to ask you a few questions'
- 'Can you please confirm your, name, date of birth and first line of your address'
- 'Have you had any previous radiotherapy?'
- 'Do you have a pacemaker?'
- 'Can you also confirm which area of the body you are expected to be treated today?'
- **Top tip:** Don't forget, if your patient is between the age of 12 and 55, you must also establish pregnancy status.
- 'What do you know about the process of radiotherapy?' This is a baseline question that can help you determine how much detail to go into. If the patient requires any pre-treatment preparation, this is a good time to go through it. Remember to also ask if the patient requires any assistance ambulating.
- You can then go through the potential side effects. Always remember to re-assure the patient that everyone is different so they may not experience any of these side effects. However, if they do, they can contact any member of staff who will offer them support or direct them to someone who can. You can also go through side effect management, such as the SOR skin care guidelines, if that is part of your practice, as well as any complementary therapy such as massage that is available to them.

- Go through the patient appointment list, all the different appointments can be confusing, so take care to point out what any special codes mean.
- 'How will you be arriving to the appointments?'. This line helps you establish whether the patient requires hospital transport or parking permits.
- Lastly, ask the patient if they have any questions and reconfirm consent.

Staff Scenarios

Often communicating with staff can be just as nerve-wracking as communicating with patients and their families and carers. Below are some conversation starters you can practice.

Asking about Breaks

If possible, try to start this conversation in a quiet period at the start of your shift with your named radiographer. Remember, do not time your break when a patient you would like to treat will be arriving.

- 'Hello (insert radiographers name)'
- 'If it will work well for the machines schedule, I would like to take my break at (insert time)'

If you are feeling unwell or have a personal emergency, you are always encouraged to ask for a break as seen below.

- 'Hello (insert radiographers name)'
- 'I am feeling unwell/exhausted/have a personal emergency, and need to take a break'
- 'I will be in (insert place) for (insert time)'

If you need to go home, let the radiographers know where you are going and if you will not be back. It is very important that a member of staff always knows your whereabouts in case of an emergency.

Asking a Question about a Technique and/or Image Match

'I would like to ask you a question about (insert technique)'

'I know you are quite busy so could you let me know an appropriate time to discuss this?'

'In the meantime, can you please direct me to the (insert resource you require) so that I can research/write down/reflect upon the technique'

If you would find it helpful for the radiographer to go through the steps, or you would like to, as it is happening, you can adapt this script.

'Hello (insert radiographers name)'

'I would find it really helpful if you/or me could talk through the steps as you are doing them'

Remember, if a patient is present, communicate what will be happening and why and check that they are comfortable with this. You can communicate this by saying the following.

'Hello (inserts patients name)'

'As part of my learning, would you be comfortable if (insert name) talks aloud the steps of this set-up'

STUDY RESOURCES

Utilise the resources below to help you study and consolidate your learning. The template below (Table 9.2) can be adapted to any anatomical site of your choosing and can be very helpful regarding any coursework based around a case study or exam.

Oncological Management Templates

Anatomical Site

TABLE 9.2 Revision Table 1 – Anatomical Site

Diagnosis	**Staging**	**Grading**	**Symptomology**	**Investigations**

Oncological Management Options (Fill Out the Relevant Boxes)

TABLE 9.3 Revision Table 2 – Oncological Management Options

Full Prescription	**Chemotherapy**	**Surgery**	**Hormone Therapy**	**Radiotherapy**	**Radiotherapy Technique**

Pre-treatment: Fill Out the Relevant Boxes

TABLE 9.4 Revision Table 3 – Pre-Treatment

CT Scanning Levels	**Immobilisation**	**Patient Position**	**Pre-treatment Preparation**	**Contrast**	**Permanent Mark Position**
Superior:					
Inferior:					

Planning Target Volumes

- Identify what is included in:
  - GTV
  - CTV
  - PTV
- List the routes of spread for each cancer type you have chosen
- List the organs at risk and their dose constraints
- Identify any side effects the patient may experience and any side effect management that is available.

TABLE 9.5 Revision Table 4 – Side Effects

Acute	Chronic	Management

QUIZ

Test your knowledge and practice your study skills with the 30 questions below. All answers can be found in this book.

1. Identify the main role of a therapeutic radiographer
2. Define radiotherapy and discuss its use in the treatment of cancer
3. List two types of radiotherapy
4. Outline the main role of a therapeutic radiographer
5. Discuss why fractionated radiotherapy was an important development for modern oncological management
6. Identify and discuss one treatment modality used in radiotherapy
7. Justify why student radiographers are important in a radiotherapy department
8. Evaluate the role of imaging in modern radiotherapy
9. Define and discuss brachytherapy
10. Outline the types of non-technical skills a student radiographer should develop
11. Identify three types of tissue present in the human body
12. Define the difference between somatic and germ cells

13. List the four stages of mitosis
14. Outline the stages of interphase in the cell cycle
15. Briefly discuss the stages of carcinogenesis
16. Identify the two main types of radiation action on cells
17. Define the acronym LET
18. Justify why DNA repair mechanisms are important during the cell cycle
19. Evaluate the five Rs of radiobiology
20. Outline the difference between stochastic and deterministic effects of radiation
21. Identify where X-rays sit on the electromagnetic spectrum
22. Define Brensstrahlung radiation
23. List three factors that affect X-ray output
24. Outline the steps in X-ray production that occur in the x-ray tube
25. Discuss the difference between characteristic and Bremsstrahlung radiation
26. Identify two main types of X-ray interaction important for diagnostic imaging
27. Justify the use of filtration in an X-ray tube
28. Define the photoelectric effect
29. Outline the clinical relevance of Compton scattering
30. Evaluate the importance of understanding X-ray interactions with matter in radiotherapy

Glossary

TREATMENT

CNS	Central nervous system
DIBH	Deep inspiration breath hold
EBRT	External beam radiotherapy
EEBH	End expiration breath hold
FSD	Focus to skin distance
HDR	High dose rate brachytherapy
IMRT	Intensity modulated radiotherapy
LDR	Low dose rate brachytherapy
MRL	Magnetic resonance linac
MU	Monitor units
OAR	Organ at risk
RT	Radiotherapy
RTOG	Radiation therapy oncology group (toxicity grading)
SBRT	Stereotactic body radiotherapy
SRS	Stereotactic radiosurgery
SSD	Source to surface distance
VBH	Voluntary breath hold
VMAT	Volumetric arc therapy
#	Fraction

IMAGING

CBCT	Cone beam computed tomography
DRR	Digitally reconstructed radiograph
IGRT	Image guided radiotherapy
KV	Kilovoltage
MRI	Magnetic resonance imaging
MV	Megavoltage

PET-CT	Positron Emission Tomography – computed tomography
US	Ultrasound
4DCT	Four-dimensional computed tomography

PLANNING AND DOSIMETRY

cGy	Centigray
CTV	Clinical target volume
GTV	Gross tumour volume
Gy	Gray
IM	Internal margin
ITV	Internal target volume
POP	Parallel opposed pair
PRV	Planning risk volume
PTV	Planning target volume
QA	Quality assurance
TLD	Thermoluminescent dosimeter

MISCELLANEOUS

FBC	Full blood count
IPW	Inter-professional working
IV	Intravenous
MDT	Multidisciplinary team
PICC	Peripherally inserted central catheter
Px	Prescription
Tx	Treatment

Index